SHEEP

A Guide to Management

SHEEP
A GUIDE TO MANAGEMENT

EDWARD HART

THE CROWOOD PRESS

First published in 1985 by
THE CROWOOD PRESS
Crowood House, Ramsbury,
Marlborough, Wiltshire SN8 2HE

British Library Cataloguing in Publication Data

Hart, Edward
   Sheep: a guide to management.
   1. Sheep
   I. Title
   636.3'083   SF375

   ISBN 0–946284–075 (HB)
       1–85223–220–X (PPC)

Acknowledgements
ICI Agricultural Division; Michael Ryder; John Read,
1980 Shepherd of the Year, for checking proofs and adding
valuable suggestions; Alun Evans, Chairman, British Wool
Marketing Board, for writing the Foreword; my wife Audrey,
for help with photographs and her usual admirable index;
Chris Brown, for her speedy photographic work.

Typeset by Inforum Ltd, Portsmouth
Printed in Great Britain at the University Printing House, Oxford

Contents

	Introduction	1
1	Enterprises	3
2	Breeds	16
3	Autumn and Winter	34
4	Lambing	44
5	Shearing and Dipping	63
6	Equipment	76
7	Flock Health	91
8	Show and Sale	102
	Glossary	112
	Further Reading	116
	Useful Addresses	117
	Index	119

Foreword

More and more people are keeping sheep for the first time – like every other animal those sheep have to receive attention at the right time. Unless that attention is given, not only will the sheep suffer, but their owners will not be able to reap the rewards of good husbandry.

To make a commercial success of any enterprise one needs to plan and to understand the options available. The author of this book has highlighted these options, both in terms of choice of breed and choice of system, together with all the husbandry and commercial decisions facing the flock owner.

When a job is done well it gives one a great deal of satisfaction – this applies in sheep farming as well as any other occupation. Joy to a shepherd is to see sheep in good condition and thriving, and when that is achieved the commercial success of the enterprise is likely to follow.

I would like to congratulate the author in realising the need for such a book as this on sheep farming.

Alun Evans
Chairman elect of
the British Wool Marketing Board

Introduction

Sheep stocking rates in Britain vary from eight or ten ewes to the acre on very intensive systems to one ewe on eight or ten acres on the highest, barest mountains. The smallholder must obviously choose an intensive system, but too high a concentration is inadvisable for the inexperienced. There is an old saying, 'a sheep's worst enemy is another sheep'. That refers to the worm burden, and without proper control over worm numbers, sheep fail to thrive and tend to die back to the 'natural' numbers that the ground will support. Modern worming techniques have changed the situation, but its basic truth must always be borne in mind. 'Beginner's luck' applies to sheep, particularly when they are introduced to land that has been sheep-free for a year or more.

The most satisfactory sheep systems are those where clean grazing is available each year, i.e. fields that did not carry sheep in the previous grazing season; but such an ideal is not always practical. Cattle, cropping or conservation in the intervening years constitute an ideal break in the worm cycle. Best of all is the use of sheep in an arable system, where they graze one-year leys very intensively, after which the temporary grasses are ploughed up, and all worm eggs with them. For the smallholder on an intensive cropping system, it may be perfectly feasible to set aside say five acres each year, and stock them at six to eight ewes and their lambs per acre on land suitable for arable. Such a flock of thirty or forty ewes can contribute materially to income, and leave the land in better condition for cropping. Stocking rates are always reckoned in terms of ewes per acre, the lambs not being brought into the reckoning until they are weaned. On many smallholdings, particularly in upland regions, such alternate husbandry is not feasible. Sheep may still be grazed intensively on the same ground year after year, but much more care with worming routines is needed, and an eagle eye kept open for signs of trouble indicated by dirty tails in the lambs.

Shepherding today consists of such day-to-day observation, coupled with the application of modern business techniques. Help is at hand on

Introduction

Island scene; Clun Forests on the Isle of Man.

all such matters, and the best move a sheep farmer can make is to join the National Sheep Association, and make contact with local officers of the Meat and Livestock Commission, Agricultural Development and Advisory Services, or the Scottish or Welsh colleges as the case may be.

1 Enterprises

CLASS OF SHEEP

So many options are open to the professional sheep keeper that it is as well to define them. Acreage, soil and climate, housing and markets available, and the aptitude and desires of the owner must all be taken into account. These options include:

Early fat lamb
Mid-season fat lamb
Selling store sheep
Producing breeding stock for sale
Breeding from ewe lambs
Pedigree stock
Rare breeds
'Gimmering'
Feeding store lambs
Wool and pelts
Milk and cheese

The question of cash flow is also important. If a quick return is essential, such operations as early fat lamb or feeding of store lambs take precedence over longer-term projects like production of breeding or pedigree stock.

Early Fat Lamb

This is a quick turnover venture without too high a capital need. Except in very favoured areas, housing is required. The ewes are generally Suffolk-cross, Dorset Horn or Polled Dorset, though other halfbreds may do, particularly if artificial means are to be used to induce early oestrus.

Older ewes that have been culled from other systems may be suitable. 'Teaser' or vasectomised rams help bring them in season. Bought in July, they are tupped to lamb from December onwards, and their offspring marketed for the Easter trade, or to catch the period when EEC prices are high. The dams may be sold in the same period, and the process repeated the next season. Not everyone approves of this system of grading the lambs after only ten or twelve weeks of life, especially on a small farm where animals are known individually. But for those with suitable accommodation, it is a paying proposition, with the further advantage that second rate pasture suffices. During the period of peak nutritional demand, the ewe is housed and fed intensively anyway.

Mid-season Fat Lamb

This is one of the easier options. For that reason, guaranteed fat lamb prices are at their lowest, so production costs must be kept right down. Ewes lamb to the grass, and so require the minimum concentrates after lambing. They are tupped in late October or early November, the optimum time for a high lambing percentage in most breeds. Either halfbred or draft hill ewes are commonly used for this purpose. Once established, flock owners often replenish through the purchase of ewe lambs each autumn, and keep them for as long as they will wear, i.e. to five or six years old, or even longer.

To start with, it is probably better to buy older ewes, rather than have both sheep and shepherd in the novice category. Then, with one or two lambings over, a younger class of sheep is bought from the profits on the old ones. Choice of ewe is wide, at both lowland and upland autumn markets.

It is essential to have good grass capable of fattening as many lambs as possible off their dams, and the remainder on saved grazing by late September. After that date, feeding costs tend to rise, although the end price is also improving.

Mid-season fat lamb really fits in better with large-scale operations, when numbers and ease of management offset the lower returns per sheep. The smallholder may prefer a system bringing more cash for the end product, but it is a good method when experience is being sought, for the shepherd is working with nature's calendar rather than against it, and selling fat lamb from July to October may well fit in with the rest of the farm.

Selling Store Sheep

A variant of mid-season fat lamb is to run the crop on and sell as stores in early autumn. There is then no need to provide pastures good enough for fattening, or much supplementary feed. The store lamb is one that is weaned from its mother, and able to graze and put on weight independently. It should be healthy and with a good frame, but carries insufficient flesh to pass the grader. It needs weeks or months of better keep to accomplish this.

In some years, the store price exceeds the grade price, but only if there is an abundance of lowland keep, so that farmers with grass to

Fat lamb production in lowland Northumberland.

Enterprises   *Breeds*

spare can afford to speculate, knowing that grade prices will rise steadily for some months. A poor national lamb crop and a shortage of store lambs on the market has the same effect.

The smallholder is at a disadvantage in that the best store lamb prices go to large, level lots that the arable or mixed farmer can finish without complications. Small pens of mixed sizes are less attractive; the little lamb always seems to stand out and drag down the price for the rest.

Producing Breeding Stock for Sale

Several facets come under this heading, ranging from pedigree stock, both male and female, to the presentation of halfbred females suitable for fat lamb production. The latter practice is well established, and consists of buying draft ewes from the hills at the autumn sales, mating them with a sire of one of the approved crossing breeds, and selling the ewe lambs in the following autumn.

Choice of draft ewe depends largely on district. A Welshman naturally goes for the Welsh Mountain, Beulah or Radnor, a north of England breeder for the Dalesbred, Rough Fell, Lonk, Scots Blackface or Swaledale, while a Scot finds ample choice among both Northcountry and Hill Cheviot, Blackface, or Swaledale in some areas.

Any old ewe and any young tup will produce lambs of a sort, but the provision of halfbred females for sale is not as simple as that. The ewes must be of good type, with suitable coats, and of contrasting face and leg colour in the dark-faced horned breeds. Buy them all from one flock and they are likely to settle better in their new home. Odd lots of hill ewes may tramp round and round the fence, pining for their familiar surroundings. A number that know each other adapt more readily to the change.

Buying the necessary crossing ram is a lottery. No one can really foretell the colour of lambs that will be left by a Blue-faced Leicester, Teeswater or Wensleydale ram without a progeny test. To reduce the odds, it is necessary to buy from an established flock, and that is not cheap.

COSTS

Prices for these crossing sires often run well into three figures, and are

6

Hill sheep on Mull.

generally higher than for Down rams. Older rams which throw lambs with bonny faces seldom come onto the market except at dispersals, and are even more expensive. The small breeder is at a further disadvantage, for if he has only twenty-five ewes, his cost of service per ewe is twice that for the flock of fifty.

In 1983 there was a collapse in the Blue-faced Leicester market, but this may not recur, as some who had greatly increased their flock sizes reduced accordingly. It is quite feasible for the smallholder to breed his own crossing sires, and many do. One or two ewes of the chosen breed, perhaps sent to a friend's ram, should provide the necessary sheep.

Out of season lambing. Shepherd of the Year Jack Thomas with his Polled Dorset tups. The flock lambs in October and November.

BUYING DRAFT EWES

Returns from the halfbred ewe lamb have been so much better in the early 1980s that hill farmers have tended to breed some themselves, which has led to a shortage of suitable draft ewes, and a consequent high price. Essentials are a fairly roomy ewe to carry one or preferably two big lambs to the crossing ram, and of the required colour according to breed. Though it has been proved at Redesdale Experimental Husbandry Farm, Otterburn, Northumberland, that light-faced Mules (Blue-faced Leicester × Swaledale or Blackface) are just as productive as the bonny dark or mottled ones, attractive colour sells. The breeder's prime object is to make money from his sheep, not to try and educate the industry!

If forty draft ewes are kept and lamb 150 per cent, then thirty ewe lambs should on average be born. Allowing for accidents and sub-standard sheep, twenty-five may reasonably be expected for sale presentation. The other half, the wethers, are sold either as stores or fat, and must not be neglected. On average, half the lambs born will always be males, a fact to be borne in mind when selecting breeding stock.

One rule of thumb is not to pay more for the draft ewe than may be expected from the sale of her first lamb crop. Thus, if the hill ewe cost £60, and averaged a lamb and a half, four would cost £240 and rear six lambs. To meet the criterion, the three wether lambs would have to average £30 and the three ewe lambs £50, i.e. £90 + £150 = £240. Matters don't work out just like that, but variations may be made if ewe price and lambing percentage are lower. The cost of the draft ewe is a major factor, and the longer she lasts, the less the depreciation. On a small farm, individual attention may be given to older ewes to prolong their productive life in a way impractical with a large flock.

VARIANT

A variant of this system is to retain the halfbred ewe lambs, breed from them twice and then sell them as two-shears or four-tooths. Tupped in their first year, they should give a 50–75 per cent lamb crop, average at least 150 per cent as shearlings or two-tooths, and then be sold that autumn. They are at maximum value, proven mothers, and exactly what the large-scale commercial producer seeks. The small farmer who spends time and care lambing young, inexperienced sheep, is rewarded for this at the sale, and saves on the number of draft ewes to be bought in.

Pedigree Stock

This is a small farm possibility, for feed costs per sheep are the same, but sales potentially higher. It is not a line for the novice, however. Pedigree stock breeding is a hard, professional, competitive game. Unlike the fat market, there is no guaranteed outlet. The first essential is to consider where the sheep will be sold. It would be quite futile for someone in Essex to start with Border Leicesters, or with Exmoor Horns in Scotland. The market simply isn't there, and the new starter cannot afford to build it. There is always a market for Down rams, but it is well supplied, and a popular breed like the Suffolk is indeed difficult to break into. A start with an in-lamb ewe of one's favourite breed is always a possibility, but not as a main source of income.

Rare Breeds

A recent opportunity concerns rare breeds. These are now sought by the numerous country parks wishing to provide interesting and unusual animals for visitors. The smallholder is well placed to supply such a need, as numbers are small, and matching pens are not required. A visit to the Rare Breeds Survival Trust show and sale at the National Agricultural Centre, Stoneleigh, Warwickshire, is advised. If that is impractical, RBST, Winkleigh, Devon, will supply catalogues and prices. A wide range of breeds fall into the rare category, from large Leicesters and Lincolns to the tiny Hebridean sheep (*see under* Breeds).

Ewe Lambs to Gimmers

A number of options are open to those who want to keep sheep without a lambing time. 'Gimmering', as it is termed in the north of England and the Borders, consists of buying in a good type of ewe lamb at the autumn sales, wintering, taking a clip, and selling after a summer's grazing as a maiden shearling. Such animals may be heavily stocked, do not demand the best pasture, and are comparatively worm-resistant. With average management and luck, losses should be few. Financial results depend largely on how well the sheep are bought and sold. In steady times there is a margin of £20 or more between the two ages, but farming is seldom steady for very long. A buoyant lamb trade may be followed by a slump the next year, when the shearlings or two-tooths bring no more than was paid for them twelve months previously, in which case they have been kept for a year for no more than the fleece. However, in these circumstances ewe lambs would be comparatively cheap to buy in, so the cycle tends to balance itself. The risk must be borne in mind, the best insurance being to buy only quality ewe lambs which usually find a satisfactory market even in depressed seasons.

Feeding Store Lambs

Another system depending largely on buying and selling price, and marketing skills, is the feeding of bought-in store lambs. Breed is immaterial. 'Anything that's cheap' is the watchword of several successful exponents, but they are experienced technicians. The beginner can easily make a mistake. Nevertheless, a heavy crop of turnips or swedes, or clean sugar beet tops, can carry over 100 sheep per acre for a limited period, the time obviously depending on the weight of crop. Halfbred wether lambs and Down crosses are usually bought, and in preparing a budget, don't forget bank interest. Physical risks are small, and with local advice on the most suitable lambs for the soil and crop, a useful profit is possible. Cull ewes also serve for the purpose; costing less than lambs and selling for less, they have made their growth, and merely need rounding off.

Wool and Pelts

In the southern hemisphere, sheep are often kept purely for their wool, but that seldom applies in Britain. Our fat lamb market is too buoyant by comparison. However, sheep people should be aware of the possibilities, of the fact that farming moves in cycles, and that Britain's prosperity was once based on wool.

The *Merino* is the chief wool breed. A system is being developed in Australia of keeping Merino wethers in slatted pens, free from stress,

11

where most of their feed is converted into clean wool. Acreage does not count in such instances. I personally would prefer to cross a hill burn, crook in hand and dogs at heel, but the aspiring sheep farmer must keep abreast of developments, especially those with scope for intensification.

The two chief exponents of Merino breeding in Britain are Dr Michael Ryder, Animal Breeding Research Organisation, Roslin, Midlothian, and Mr Geoffrey Boaz, Rhydd Farm, Hanley Castle, Worcester. The British Wool Marketing Board is not fully involved, nor is its price schedule very encouraging for Merino producers.

In Sweden, sheep are kept for their pelts, and a breed has been developed for the purpose. Combined with home curing, the possibilities are worth exploring, but there is no guaranteed end price as in the case of lamb.

Root fattening for Mashams and Suffolk crosses.

The milking flock; British Friesland ewes heading for the parlour on a Dumfriesshire farm.

Milk and Cheese

This new branch of sheep production offers real scope for smallholders. Cheese making from ewe's milk is catching on, thanks to the drive and enterprise of a few pioneers. Strictly speaking, it's not new, but rather a revival of a traditional form of husbandry. Monks and shepherds made cheese from ewe's milk in medieval times, and sheep houses were built for the purpose on the North Yorkshire Moors and other places, but unfortunately very little of the layout remains. The practice has continued without a break for centuries on the Continent, and the breed chiefly used in the British revival is of Dutch origin. The British Friesland yields some 125 gallons in a lactation lasting eight to ten months, and half a gallon of milk makes 1lb of cheese. As soft cheese is sold wholesale at £1 per lb, and hard cheese at £2, the gross output could be £250 per ewe, plus the value of fleece and lambs.

This is potentially encouraging, but the bare facts hide a lot of hard work. As with a dairy herd, the sheep must be milked twice a day, every day, but the bulk milk tank is not the end of the matter in the ewes' case.

13

Enterprises

Cheese making demands an equally careful routine, and the finished product requires two or three months to mature. There is no guaranteed market, but sufficient outlets are being found by existing breeders to cater for flock expansion.

BREEDING STOCK

Prices for breeding stock are £150 to £200 apiece, either male or female. The breed is prolific, with a lambing percentage exceeding 200. March-born ewe lambs can reproduce themselves in the following January. One Scottish farmer and his wife run fifty British Frieslands on sixteen acres which includes reclaimed ground. They have now increased their enterprise. They give one warning; the breed is susceptible to copper poisoning, and it is not easy to find a dairy-type concentrate without copper.

The milk is too valuable for lamb rearing. An infra-red lamp is used to dry the new-born lambs, which are given 4oz (113 grams) of colostrum in two feeds separated by four hours, and then reared on ewe milk substitute. Milking is by Fullwood machines. They have a PVC liner of a type widely used in Europe, a pulsation rate of 120, and a vacuum of thirteen; some French flocks use eleven. The ewes are milked in an eight-abreast parlour, being held by a rail which is lowered as they put their heads down to eat. They are even more idiosyncratic than cows, preferring a particular batch of companions or a certain stall. They have a double let-down of milk, which does not complicate the milking process unduly provided there are plenty of stalls. Eight ewes can be milked in twelve minutes, including movement in and out. Conquering the techniques of cheese making is a vital part of any success.

Profit Factors

Three main performance factors influence flock profitability:

1. Number of lambs reared.
2. Stocking rate.
3. Value of lambs sold.

Stocking rate accounted for 40 per cent of the superiority of top third

14

flocks in Meat and Livestock Commission statistics on lowland flocks, lambs reared 17 per cent and lamb value 11 per cent. There are wide variations in the seasonal guide price – the sheep farmer's guarantee – which change from year to year and must be studied. Turnover also counts, e.g. feeding two medium lambs in succession rather than one large one. Prices are usually lowest in July–October, and highest February–May.

Ram Breed Comparisons

Carcass weight and the rate at which lambs mature to the desired fat levels are directly influenced by breed. It is far easier and cheaper to change the breed of ram than to change the ewe stock. The Meat and Livestock Commission conducted controlled trials over five years at nine locations, using ten ram breeds on three breeds of ewe. Relationship between carcass weight and fatness is important, and in MLC's 3L Fat Class, Southdown-sired lambs averaged 16½kg (36lb), Dorset Down and Hampshire 17½kg (38lb), Suffolk and Texel 19½kg (43lb) and Oxford 20kg (44lb).

The chief lesson for the novice or small-scale sheep keeper is to recognise the importance of statistical information, to approach the sheep enterprise with 'dairying-type precision', and to plan and adhere to a budget while in no way neglecting the husbandry aspects.

2 Breeds

MARKET

There is no 'best breed'. Each has its place, and no one would try to establish a Romney flock on the slopes of Snowdon, or Blue-faced Leicesters at 2,500ft on a Lakeland fell. Equally important as suitability to environment is sales practice. The smallholder keeping sheep for profit must have a clear end in view. He is there to make money, not to convert others to his way of thinking or to tread the path of the pioneer. These laudable activities are too expensive for the small-scale breeder making a living from stock. In some areas there is a prejudice against white faced sheep; in others, butchers require lambs of a certain weight. The customer is always right, and it is the breeder's job to supply what is wanted. We find that, though Welsh Mountain ewes thrive in the Scottish Borders, no one is interested in buying a Welsh ram there.

CHOICE OF BREED

The most important aspect is to fit the sheep to the ground. Any attempt to reverse this principle is doomed to failure, as has been proven many times. Britain's many breeds have evolved over the years to suit a particular environment, though the reasons why they do so are sometimes obscure. Reaction to trace elements and the ability to absorb copper could be one factor; others are the way in which high rainfall or soft ground are withstood. Lowland breeds need a greater supply of feed at most times of the year than do hill breeds.

Local Types

A sound rule for the beginner is to choose a breed already popular in the locality. They have been found to suit the soil and climate over the

16

years, there is a recognised market for them, and they may be bought without too high a transport cost. If one is establishing a flock in an area where sheep are few, such as intensive arable or dairying districts, then a safe guide is to go for a breed or cross with a national reputation, such as the Masham, Mule, or Scots or Welsh Halfbred. The Down breeds are adapted to most lowland situations, and the Clun is a good grassland sheep on better pastures.

The general rule when buying sheep is to move them from the north and west to the east and south, and from hill to lowland. In other words, give them a change for the better.

The main British and recently imported European breeds are listed here, but for the commercial smallholder, the various crosses may be of higher importance. A number of breeds are kept almost entirely to breed rams to sire those crosses. This aspect has attractions for the smallholder, as little more feed or acreage is required to keep stock whose potential sale value is well above commercial rates. Such sheep will cost more in the first place, and a market must be found. A lamb may be graded at any centre provided it has the requisite fat cover; and it will then attract the Guide Level price even if the market is slack. The vendor of breeding stock has no such market floor. To gain a premium, his name counts, and the process takes some years of expense and effort. Nor is the sale of breeding stock for the unskilled. Any Blue-faced Leicester ram put to any Swaledale or Blackface ewes will produce Mules, but they will not be of the quality to attract a premium over fat price unless both parents are carefully selected, and even then, chance plays a part.

Down Breeds

The Down breeds and the Ryeland are meat sires. They are seldom kept pure for any other reason than to provide crossing rams to put on milky ewes, and all their crossbred progeny is, with a few exceptions, destined for the table.

RYELAND

The Ryeland is one of our oldest breeds. It is small and neat, sturdy at birth and quick maturing. It is a possibility on early lambing halfbred ewes whose progeny is required to be sold before the low price trough starts in July.

SUFFOLK

The Suffolk is the most popular Down breed. High prices are made, and there is a wide market. Like other Downs, the Suffolk is not very prolific, but for those who fancy breeding it pure, the best advice is to keep a ewe or two to provide rams for one's own halfbred flock, and ease one's way into selling surplus breeding stock. Suffolks are bred by highly professional people with large flocks, and the smallholder may prefer to venture where competition is less severe. An important attribute is the Suffolk's propensity for early lambing, whether pure or crossed.

DORSET DOWN

The Dorset Down (not to be confused with the Dorset Horn or the Polled Dorset) is another prime lamb sire with good export record. Its main breed show is at Taunton in July, with others in the south, and also at Builth Wells and Kelso.

HAMPSHIRE

The Hampshire has face and ears of rich dark brown, and the classical, solid block of meat on four short legs symmetry of the ideal meat sheep. The breed does well on grass or arable, and when it is necessary to get lambs away early. That leaves the way clear for a second crop of bought-in stores if desired, and may pay better than carrying each sheep on to maximum potential value. Hampshires stage their breed sale at the Royal Show in early July.

SHROPSHIRE

The Shropshire declined in numbers, but has staged a come-back in which its genuine properties have been appreciated. Hardier than most Down breeds, it is also rather more prolific and clips a heavy fleece. Its carcass is lean and of excellent flavour. The main breed sale is at the

A winning Suffolk at the East of England Show.

Royal in July, but it is also found at major sheep sales in the Midlands and south.

OXFORD

The Oxford is the largest Down breed. It lost favour when the demand for small joints was at its height, but has rightly returned to meet the 'supermarket era' and new cutting techniques. 'Big tups and little ewes' was Kenneth Russell's recipe for making money from sheep when he was Principal of the Royal College at Cirencester, and the Oxford fits into that criterion admirably, with mature rams weighing well over 200lb (91kg).

Oxfords have largely forsaken their home county, and several top flocks are now found in Yorkshire, with classes at the Great Yorkshire, and sales at Malton, Northampton, Kelso and Builth Wells. Flocks generally are small, and the smallholder could do worse than investigate this breed's possibilities, provided his ground is productive enough to match their large appetites.

Breeds

*Oxford Down ewes and lambs. Big, quiet sheep capable of
heavy stocking, but with demanding appetites.*

SOUTHDOWN

The Southdown is another for consideration. Its small, neat conformation makes it the ideal ram for maiden ewes, corresponding with the
Angus bull frequently used on dairy heifers. Welsh Halfbreds are often
bred from as ewe lambs, and the Southdown makes the ideal ram. The
lambs are small at birth, and do not reach large weaning weights, so
their demands on the immature dam are minimal during both gestation
and lactation.

For early maturity the Southdown is unbeatable; it sires most of New
Zealand's Canterbury Lamb. Breeder Hugh Clark, from Newmarket in
Suffolk, says the only way he can make money out of sheep is to get rid
of them quickly, stressing that profit comes from rapid turnover rather
than maximum returns per individual sheep. Sales are at Findon in
Sussex, Ashford in Kent, and Maidstone in Kent.

WILTSHIRE HORN

The Wiltshire Horn is unique in being solely a mutton animal, with no fleece. Its devotees claim that all feed goes into liveweight gain, that much time is saved as no dipping, crutching or shearing is needed, and that the sheep's condition may be readily assessed at all times. Another claim is that eighty-five per cent of shepherding costs are attributable to wool growing. On Anglesey, Iolo Owen has developed a commercial woolless breed based on his Wiltshire Horns, and Anglesey and Northampton are the main sale centres. The breed has great potential in a suitable climate.

OTHER TERMINAL SIRES

The Meatlinc is a terminal crossing breed developed in Lincolnshire by Henry Fell. The Meatlinc is also an amalgam, pioneered by Sheep

Texel cross ewes with Suffolk cross lambs.

Systems Promotion with the object of providing the most effective terminal sires.

The Bleu de Maine and Charollais are recent introductions claiming advantages as fat lamb sires with high killing-out percentages. The Oldenberg is a quiet sheep, originating on German marshlands, and is used as a terminal sire, as is the Texel. Most popular of the European imports, the Texel is a white-faced, white-woolled breed with mutton qualities that have brought for it some very high prices.

HALFBRED SIRES

An important group of breeds developed solely for use on hill ewes to provide the various halfbreds holds potential for the smallholder. They are best regarded as a sideline, however, which might give a welcome bonus.

The breeds are the Blue-faced Leicester, Border Leicester, Tees-water and Wensleydale. My placings are strictly alphabetical, for few topics arouse such arguments among sheep people as the best crossing sire!

BLUE-FACED LEICESTER

The Blue-faced Leicester has been developed in the uplands of northern England to cross onto Blackface and Swaledale ewes. The result is the popular Mule. Blue-faced Leicesters have a reputation of desiring to return to their Maker at the earliest possible moment, as our vet euphemistically puts it, and the shepherd's unenviable job is to circumvent those desires.

Highly prolific, the Blue-faced Leicester may have four or even five lambs, all dead. She may have a huge single, she may have twins and turn against one of them. For those with patience, shepherding skills and luck, she may pay real dividends. But don't count on anything until the auctioneer's hammer drops.

BORDER LEICESTER

The Border Leicester is slightly more sturdy, but a difficult sheep to breed correctly. A good one is among our most handsome beasts but, as with all Roman-nosed sheep, great care is needed to keep the teeth right. The Border Leicester is sire of the Scots Halfbred, out of a Cheviot; of the Greyface, out of a Scots Blackface; of the Welsh Halfbred, out of the Welsh; and of the so-called English Halfbred out of the Clun. Registrations are strict, and the Border Leicester is backed by a century of sound organisation.

TEESWATER AND WENSLEYDALE

The Teeswater and the Wensleydale are both sires of the Masham, out of Swaledale or Dalesbred ewes, and the Kendal Masham out of Rough Fells. Long and lean in carcass, and clipping a heavy fleece of lustre wool, Teeswaters and Wensleydales have stretched their boundaries beyond the northern dales, and are found fairly widely throughout Britain.

A group of Teeswater ewes – sire breed of the Masham.

Breeds

COLBRED

The Colbred has a parallel function, and incorporates genes from the deep-milking Dutch East Friesland. It is less readily available in the market.

BRITISH MILKSHEEP

The British Milksheep is based on the Fries Melkschaap, imported during the 1960s from Holland and West Germany. Average yield there is 140 gallons (636 litres) per lactation at six to seven per cent butterfat. Hybrid offspring achieved good results in trials against other halfbreds. The Milksheep is large-framed and polled, with face and legs white and clear of wool. Claimed prolificacy is 275 per cent, and both sexes breed readily in their first year.

Hill Breeds

Britain's wide range of hill breeds are of interest for two reasons. In an upland situation they may be bred pure, and on the lowlands their draft ewes may be kept for a number of years.

CHEVIOT

The Cheviot is a white-faced, white-woolled sheep originating in the Scottish Border hills of that name, and now divided into two breeds, and rather more types. The Hill (formerly South Country) Cheviot is the smaller, with prick ears, and arguably the bonniest lambs of any breed. It is worth considering as it is not too dear. It has excellent carcass qualities, and as a 'spring sheep' is at its best in the high-price January to April period. To the Border Leicester it breeds the Scottish Halfbred, and is widely crossed with a Suffolk or other Down ram for top quality fat lamb. The Hill Cheviot's drawbacks are a fairly low lambing percentage, and a propensity to get fast on its back – 'cowped' or 'rig-welted'. That is due to its wide, level back, a reminder that desirable qualities in livestock usually have corresponding snags.

The Northcountry Cheviot is extensively bred in the far north of Scotland. It is bigger, more prolific and deeper milking than its hill cousin, with a correspondingly larger appetite. A sheep of high potential, its faults are premature shedding of wool and teeth, and udder troubles, the latter associated with high milk yield.

SCOTTISH BLACKFACE

The Scottish Blackface is a medium-sized sheep found in great quantities on the hills of Scotland, northern England and the south-west. In recent years its draft ewe prices have dropped, yet it makes an admirable sheep for fat or store lamb production.

SWALEDALE

From Lakeland to the North Sea coast, the Swaledale is the most popular hill breed. As dam of the Mule, it is much in demand. For an

A magnificent Hill Cheviot ram, Inter-breed champion at the Border Union Show, Kelso, 1984.

upland breeder contemplating pure breeding, it has much to recommend it, but the buyer of draft ewes is up against such stiff competition that he may make more profit elsewhere, certainly until experience has been gained.

DALESBRED

The Dalesbred has similar characteristics; its fleece is heavier and with more curl, and experts regard it as the ideal dam of the Masham.

ROUGH FELL

The Rough Fell is concentrated around Kendal, Cumbria. Its halfbred offspring are not as fashionably marked as the other halfbreds, but it is

Swaledale rams; this breed is spreading from northern England to many hill areas.

worth considering for its docile nature and heavy fleece. It is said to be the only hill breed for which dogs are not essential.

DERBYSHIRE GRITSTONE

The Derbyshire Gritstone is based in the southern Pennines. The Gritstone has a mottled face, and hardiness compatible with the tough districts where it thrives when others would fail. It is hornless, and has been used to breed the horns off the Blackface and Swaledale in areas troubled by sheep head fly.

LONK

The Lonk is also based in the southern Pennines, and is large for a hill breed, with a dense fleece that dries quickly. One farm where it thrives on the high Pennines endures 200 rainy days a year. It is a self-sufficient sheep; it has to be, for its grazings are often fog-bound.

HERDWICK

The Herdwick is part of Lakeland. At the National Sheep Association's 1984 Cumbrian event, Herdwicks with admirable cross-bred lambs were on view. The ewes had cost £15 apiece the previous autumn, the sort of figure that makes sense to the buyer. Herdwicks are small, born black, and turn grey with age. A smallholder is unlikely to keep them pure; they are a breed of the high hills and wide open spaces where a more productive type cannot live. But put to a Suffolk or Cheviot ram, they have excellent lambs for the grade or the store trade. They live to a great age, but their fleeces are the least value on the British Wool Marketing Board's list.

WELSH MOUNTAIN

The Welsh Mountain is the largest in terms of ewe population in the UK. It weighs only 70 to 85lb (32–38kg) on the hills, and this little sheep dots the better Welsh pastures as thick as hawthorn blossom. A Welsh ewe may well have halfbred twins each almost her size by weaning. She is a kind mother, with a dense and quite valuable fleece. Lambing 90 per

Herdwicks heading for the hills.

A fine Welsh Mountain ram.

28

cent or more on the hill, the breed approaches 150 per cent on good ground. Almost half a million Welsh ewes are sold as drafts yearly, so there is ample choice, at prices commonly below the bigger hill breeds. The breed society runs a co-operative flock improvement scheme, with ram performance and progeny testing. Many types exist among the mountains of Wales, and a notable development is the Brecknock Cheviot. This has more size and scale without loss of hardiness, and is centred around those deceptively harsh hills, the Brecon Beacons.

WELSH HILL SPECKLED FACE
The Welsh Hill Speckled Face is a hill breed of mid-Wales, rather larger than the Welsh Mountain.

Half Breeds
The crossing breeds and the hill breeds on which they are used have been considered. The main resulting halfbreds are as follows.

MASHAM *(Teeswater or Wensleydale × Swaledale or Dalesbred or Rough Fell)*

Giving a heavier clip than most, Mashams are also first-class butchers' sheep. As half of all lambs born will be males, this is important. Mashams have many times featured in the greatly missed National Lambing Competitions, and many sound farmers prefer them to Mules. They are excellent mothers and milkers, and are usually put to a Down ram.

MULE *(Blue-faced Leicester × Blackface or Swaledale)*

Today's most popular fat lamb dam, the Mule ewe shows kindness and concern for her offspring. In large flocks this is vital, and it is important in the small flock where the owner is shepherd and has a score of other jobs on hand. Winners of *Livestock Farming/Rumenco* Shepherd of the Year Award have frequently herded Mules. A true lambing average of 170 to 190 lambs reared per 100 ewes put to the tup is frequently achieved from Mules. Such prolificacy makes them dear to buy, but remember that it has been proved at Redesdale Experimental

Husbandry Farm, Otterburn, Northumberland, that light faced Mules perform every bit as well as the fashionably expensive bonny mottled faces. The North of England Mule Sheep Association was formed in 1980.

WELSH HALFBRED *(Border Leicester × Welsh)*

Smaller and less prolific than the first two, this breed earns consideration. It is frequently bred from as a ewe lamb, wears well, is a kind mother, may be heavily stocked and costs less to buy. Its well-run official sales offer enormous scope, and are backed by a field service that ensures satisfied customers. Its organisation is twenty-five years ahead of the others.

WELSH MULE *(Blue-faced Leicester × Welsh Mountain, Welsh Hill Speckled-face, Beulah Speckled-face, or Brecknock Cheviot)*

Following its formation in 1979, the breed association has successfully promoted the main sales, Welshpool and Builth Wells. Like the Welsh Halfbreds, its stock is inspected before being allowed on sale.

SCOTS HALFBRED *(Border Leicester × Cheviot)*

For years the leader in the field, this is bigger than the other halfbreds, though the Hill Cheviot cross is more compact. A breed for the fertile arable and mixed farming areas, it has a large appetite but a capacity for very satisfactory performance under the right conditions. A lambing percentage of 166 is frequently attained. Many flocks are successfully in-wintered, and its kind, placid nature suits it to intensive systems.

SCOTTISH GREYFACE *(Border Leicester × Blackface)*

Among the first halfbreds, and still a good one. In fact, its fleece type and face colour are not highly regarded at the moment, and as neither is vital in fat or store lamb economics, there is much to be said for the Greyface for novices. Less money is at stake, performance is good, and there is a wide choice.

Masham gimmers – the bonny breed.

Welsh Halfbred ewes in North Wales. This Penbedw flock has averaged 170 per cent lambs reared over many years.

Scottish Halfbreds. A traditional cross, prolific, and good mothers.

ENGLISH HALFBRED *(Border Leicester × Clun Forest)*

A larger sheep suitable for intensive management. Lambs to a Down ram may be taken on to heavy weights without excess fat.

Grassland Sheep

These breeds are suitable either for pure breeding or crossing on lowland or lower hill conditions. Foremost is the Clun Forest, which in the hands of capable breeders has improved in conformation and lambing percentage in the past decade. Others are the Kerry Hill, from the same Welsh Borders area, the Hill-Radnor, the Llanwenog of high prolificacy, the Beulah Speckled-face for better hill pastures, and the Lleyn. Highly prolific, the Lleyn was once seldom seen outside north-west Wales, but now has enthusiastic breeders elsewhere.

Breeds of the South-west

Devon and Cornwall have their own indigenous breeds, none of which has spread far afield. There seems no reason why they should not do so; the Exmoor Horn has an excellent carcass, and is a good sheep on the more fertile hills, producing fine wool. The White-faced Dartmoor is a true hill sheep, grazing up to 2,000ft in summer, though usually wintered off the moor. The Dartmoor is mainly polled, with a coat of long, curly, lustre wool, with flocks that have been established for over a century.

The Devon Closewool is a grassland sheep, producing a grand fat lamb, either pure or crossed. The fleece withstands wind and rain and is of wide commercial use. In 1977 the Devon and Cornwall Longwool was officially established through amalgamating the South Devon and the Devon Longwool. It clips a heavy fleece and the hoggs are suitable for finishing in the high-price, winter period.

Rare Breeds

Any of the rare breeds is a definite possibility for the smallholder. They have a specialised market for the many country parks being established, where large numbers of evenly matched sheep are not required. They repay study, and the best place to start is at the Rare Breeds sale at the National Agricultural Centre, Stoneleigh, Warwickshire. Among them are the Boreray, Hebridean, Manx Loghtan, North Ronaldsay, Portland, Shetland, Soay and White-faced Woodland.

Also found at the Rare Breeds sale at Stoneleigh are the Cotswold, Leicester and the Lincoln. All are longwools with very heavy fleeces, weighty carcasses, quiet natures and fairly low lambing percentages. They are regaining ground due to supermarket demand for large, lean joints. The Welsh Badger-faced is another attractive and unusual breed liked by country parks. The Black Welsh Mountain is prolific for a hill breed, small and of good quality meat. Once a rare breed, the Jacob has now become firmly established commercially. It is still liked by park owners, is prolific with 'gamey' meat, and has wool much in demand by home spinners.

3 Autumn and Winter

CALENDAR

The sheep farmer's year begins in autumn. Spring and lambing time receive accolades in the popular press, but the shortening days are when it all starts, and when a lot of hard work must be done. The breeding flock is replenished, either by home-bred or bought-in females. Rams are selected, the bulk of lambs sold and perhaps others bought. Sheep sales are the order of the day. As the season progresses, breeding pens are formed and the rams loosed.

With so many different enterprises available, it is impractical to give a detailed calendar for each. Yet the same operations are carried out in the same order, whether for a pedigree Down ram breeding flock starting to lamb in late December, a flock of Mules lambing in March, or a high hill stock where late April or even May sees the start of lambing. It applies equally to Dorset Horns or Polled Dorsets, where different sections lamb in spring and autumn.

FLOCK PREPARATION

The starting point is the shepherd's preparation of his flock for the next season. He will consider numbers: is it feasible to carry more ewes, the same number, or do cropping changes and other factors demand reduction?

Udders

Having established how many should lamb down, the shepherd goes through his flock thoroughly. The basic rule is to discard any with faulty udders, although if the price of breeding stock is high and for culls low, there may be a case for keeping an otherwise outstanding ewe, and rearing one of her lambs on the excellent milk substitutes now available.

Teeth

In a large flock, and certainly in a hill stock, only ewes with sound mouths are retained. That means that they have eight sound teeth at the front of the lower jaw, evenly worn and not projecting or gappy. Such sheep are able to ingest food effectively, and compete with their sisters in the big flock. Check jaws externally for 'lumps and bumps' caused by bad molars.

For the smallholder it is rather different. He can give special treatment to 'broken-mouthed' ewes, by penning them separately and allowing extra rations if need be. If ewes have given several years' good service before their teeth go, there is a case for retaining them and 'nursing' them through the winter till the grass comes once more. Sheep with two or three projecting teeth are worse off than those with no incisors left, but it is a risk. Keeping old ewes with records known to the

Sound mouth on a Swaledale ram – incisor teeth should just meet the pad.

shepherd, and accustomed to the place and the routine, is different from buying a batch of someone else's cast-offs, a practice not recommended.

General Health

Others to make the market trip are ewes subject to prolapse, or which have failed to regain condition after weaning, or have proved barren for two years running. Any in the last group should be sold in winter or spring, when prices are high. A ewe missing just one year should be given a second chance, but not more.

There is also a case for selling persistent fence breakers. Persistently lame sheep may be discarded; very occasionally it proves impossible to cure an individual of foot troubles, though the fault was probably due to bad shepherding at some early stage. A ewe that was a thoroughly bad mother may also go, a fate she will have been promised since last lambing, but a young sheep should be given chance to redeem herself.

Quality

Lastly, culling gives an opportunity to raise the general level. Here the shepherd is bound by finance and number of replacements available. It is very easy to decide which conform to his ideal of breed type, less so to find better replacements. In an ideal situation, more gimmers or theaves are available than ewes to be culled. The best young stock are then picked out on a basis of conformation and breed type, backed up by parents' records if available. The ewes are similarly selected, using their own records. Obviously, the higher the lambing percentage, the greater the scope for selection and breed improvement in this way. In hill flocks it is common practice to sell draft ewes at either three, four or five-crop, i.e. as four, five or six year-olds, perhaps keeping a few of the very best for ram breeding, and self-replacing lowland flocks may follow a similar system, thus selling mainly ewes of high value rather than low-price culls.

Ram Selection

A minimum of one mature ram per fifty ewes is needed, with ram lambs serving half that number. In a cross-breeding flock for fat lamb, ram selection is based on conformation, growth rate and preferably on records. There are large numbers of well-established breeders of Down rams, and the best ones are just as keen as the buyer that the rams do well. If you buy from such a flock, either privately or at a sale, then assistance is more likely to be forthcoming if the ram goes lame or proves infertile. That is one more reason for choosing a breed or cross common in the area; when accidents happen, as they will, substitutes are readily available.

The breeder of halfbred stock sold for fat lamb production is on a stickier wicket. With a ram lamb or uproven shearling he is indulging in a lottery. Only the lamb crop will prove whether the sire is good enough, and the best way to shorten the odds is to buy from a flock of high repute, which of course means paying a premium.

Pure Breeding

Any mistake made in the two categories mentioned is for one year only, but in a pure-bred flock an inferior ram can cause lowered standards for years to come. Yet the cost of the best sires may be far beyond the smallholder's purse, and this applies particularly to certain hill breeds. A stockbreeder's general principle is to choose a sire to set right a deficiency in the existing flock; if the quarters are too sloping, choose a blocky type of ram, or if flock quality is on the coarse side, go for a neat ram.

He should of course have all the breed points. How easy it is to state this, yet how difficult to go to a sale and find such an animal within the permitted budget. In such cases, settle on a few points that must be right; one Swaledale breeder always picks a ram with fleece to suit his conditions, so that he doesn't have to worry about that highly heritable factor. He chooses rams with sound teeth and legs – not always easy – and after these fundamentals goes for the ones knocked down within his price range. Too close relationship to the ewe stock should be avoided, as should any known association with scrapie.

Ram Management

As the days shorten and the leaves change colour, so the rams' characters alter. Having spent the summer dozing in the sun like old gentlemen on a park bench, they now become restless and belligerent. Examination by the shepherd or vet of the rams' reproductive organs several weeks before tupping is time well spent. Semen testing is relatively cheap, and essential if only one or two rams are used. Improvement in nutrition, either with or without concentrates, six weeks before use, gives better semen samples and tamer rams.

The ram paddock must be absolutely secure, for your own and your neighbours' sakes. An airy shed is better than a doubtful field, and gives less room for fighting. Horned rams are occasionally coupled together so that they cannot fight, but one may keep the other from feed, and particularly water, with disastrous consequences. Feet should be trimmed if necessary, with a footbath walk as routine, or failing that a precautionary squirt from an anti-footrot spray. It is amazing how rams remain sound all year and then go lame two days before needed.

Ewe Management

After weaning, the ewes are kept on moderate keep till their milk dries up. A week after weaning, check udders to ensure there are no hard lumps or other complications, and start to improve keep.

Some authorities prefer the ewes on a rising plane of nutrition as they go to the ram, others follow work done in New Zealand which points to a high body condition for some weeks previously. All agree that ewe condition must *not* be declining. Silage or hay aftermath is ideal for 'flushing', as is a green crop specially grown. Other competent stockmen make do with a few concentrate nuts.

Body Scoring

Wool is a wonderful substance. It protects the sheep against weather extremes, but it also disguises its body condition. Any sheep with dry fleece tends to look good, and on a frosty February morning all may look well with the flock. Appearances can be very deceptive, and the only way to assess a sheep's condition is by handling. A standard method has been worked out by the Hill Farming Research Organisation from Australian work. It uses a six point scale from nought to 5, though most ewes score between 1½ and 4½. Condition scoring should be used to achieve optimum flock condition at tupping (scores of 3 or 4), and as a check on condition as lambing approaches. Remedial action must then be taken; scoring is a waste of time unless followed up with extra or less feed as indicated.

METHOD

Condition scoring is assessed by finger pressure along the top and sides of the backbone, in the loin area immediately behind the last rib and above the kidneys. The Meat and Livestock Commission and other advisory bodies will demonstrate. Assessed in order are:

1. Sharpness or roundness of the spinous processes of the lumbar vertebrae (the bony points rising upwards from the back).
2. Prominence and degree of cover of the transverse processes of the vertebrae (the bone protruding from each side of the backbone).
3. Extent of muscular and fatty tissues underneath the transverse processes (pass fingers under the ends of the bones).
4. Eye muscle fullness and fat cover (press between spinous and transverse processes).

CONDITION SCORES

0 Extremely emaciated and on the point of death. It is not possible to detect any muscular or fatty tissue between the skin and the bone.
1 The spinous processes are prominent and sharp; the transverse processes are also sharp, the fingers pass easily under the ends, and it is possible to feel between each process; the loin muscles are shallow with no fat cover.
2 The spinous processes are prominent but smooth, individual processes can be felt only as fine corrugations; the transverse processes are smooth and rounder, and it is possible to pass the fingers under the ends with a little pressure; the loin muscles are of moderate depth, but have little fat cover.
3 The spinous processes have only a small elevation, are smooth and rounded, and individual bones can be felt only with pressure; the transverse processes are smooth and well covered, and firm pressure is required to feel over the ends; the loin muscles are full, and have a

moderate degree of fat cover.

4 The spinous processes can just be detected with pressure as a hard line; the ends of the transverse processes cannot be felt; the loin muscles are full and have a thick covering of fat.

5 The spinous processes cannot be detected even with firm pressure; there is a depression between the layers of fat where the spinous processes would normally be felt; the transverse processes cannot be detected; the loin muscles are very full with very thick fat cover.

Crutching

A week or so before tupping, the ewes' tails should be clipped out. Most hill breeds are left with long tails as protection and food reserve. The job is done by hand or machine, taking a couple of inches round the vulva, and a few inches of wool from the tail itself. The freshly cut wool becomes bristly for a few days, and impedes rather than aids the ram, hence the forward planning. This job may well be coupled with a foot check. Multi-vaccines are commonly given now, and worming with combined fluke-and-worm drench.

Tupping

The ewes are settled into their different lots. Then, on the strictly appointed date, each ram is taken along, and released with raddle on his chest. Do not paint the raddle too far forward, or a mark may be left if investigation rather than mating occurs. An alternative is the coloured chalk and harness, which must be fitted correctly, or it will cause sores under the armpits.

Most shepherds stay for a few minutes and watch the rams when loosed. In a flock of twenty or more ewes, the odds are that at least one will be in tupping, and an effective ram will immediately seek her out. Once he has been seen to mate successfully, the basis for an effective tupping time has been laid. In a small flock, the ewes may be noted as they are colour-marked by the ram. Or the colour may be changed after a week, so that ewes due to lamb in the first week are known.

The ewe's gestation period is 147 days, or 21 weeks, but varies slightly between individuals and between breeds. The breeding cycle is 17 days, but it will be found that not all ewes come to the ram in the first

Tail docking by rubber ring. Under present regulations, enough tail must be left to cover the vulva in females.

17 days, and three weeks is the period used by a number of top shepherds.

After 17 days the ram's colour must certainly be changed, and if a number of ewes are re-marked, indicating a return to service, then swift action is called for. A substitute ram must be found immediately, otherwise lambing will drag on for weeks.

The ram should have a daily feed of concentrates. If he has been fed nuts beforehand, he will usually come readily enough, and this training is a hallmark of a good shepherd. It is better to avoid turning up the ram to apply fresh raddle; paste smoothed onto his chest while he is eating is better. One advantage of raddle is that the sire must be caught daily, and is most easily caught by feeding. Therefore he gets his ration, come

Artificial insemination is a technique with potential for flock improvement, but has not made much impact on the sheep world as yet.

Throwing out whole Monorosa beet to Mule and Masham ewes.

42

Big bale silage – very useful sheep feed if properly made.

storm or emergencies, whereas the same does not apply to the harness, which induces lazy shepherding.

Post-tupping

The weeks following tupping are among the easiest of the year, if the weather allows. The rams may be taken out after six weeks, in which case the end of tupping is definitely known, or left longer on the basis that a late lamb is better than none. It is always a cause for comment when the 'geld' start to lamb!

Care of Ewes

In mid-pregnancy, a level condition is maintained. The ewes must not be allowed to slip back, but to feed extra is wasteful at this time. Good pasture and perhaps some hay should suffice. If a storm breaks, then more hay is fed, and on the best upland farms, it is the practice to continue hay feeding once it has started, whatever the weather.

Rough handling, severe weather or short rations during the first third of pregnancy, i.e. about seven weeks, can cause resorbtion of the foetus. As this cannot readily be detected, the ewes are classed as 'geld', when in fact poor management caused the loss. Don't use in-lamb ewes for dog training practice at any stage.

4 Lambing

BEFORE LAMBING

'The unborn lamb is out of sight for five months. Never, even for a single day, let it be out of mind.' That rule from Dr Allan Fraser is vital for the budding shepherd; if its principles are followed, the way for a successful lambing is prepared. Some people advise extra feeding only in the last six weeks of pregnancy, based on the fact that the foetus develops markedly in this period. However, inexperienced breeders should note that extra feeding at this stage may be too late.

Winter scene in Weardale, County Durham. Summer's bounty is fed to hungry sheep, whose appetites rise with cold weather.

Especially with the hill breeds, attempts to retrieve earlier under-feeding may result in too large a single lamb. I retained some very ancient Swaledale ewes which by all the rules should have been sold in autumn, and which would not keep their condition during winter. They were fed sugar beet pulp nuts and balanced concentrate at 2lb (0.9kg) a head for the last six weeks and, though all lambed successfully and had enough milk, their offspring were so large at lambing as to constitute an additional drain on the dams just when it was least needed.

Condition may be assessed only by regular, possibly weekly, handling. A sheep with dry fleece usually looks well, and the amount of flesh carried cannot be gauged except by condition scoring (*see page 39*).

Rations

The ewe's ration in mid-pregnancy is based on good quality hay or silage. Poor quality silage is useless for sheep, and if the hay is only

Feeding silage in a sheep rack.

average, it will need supplementing by ¼–½lb (113–227 grams) concentrates per head daily. Mineral licks should always be available, and housed ewes drink more and more water as gestation progresses.

Groups

Even in a small flock, some sheep will be much fitter than others. If the tups were raddled, the ewe's colour mark indicates her approximate lambing date, and eight to ten weeks before lambing the leanest ewes (condition score below 2½) should be drawn into a separate field or pen, and fed extra (½–¾lb, or 227–340 grams) concentrates.

Remaining sheep are divided between early lambers in good condition, and late lambers. Any overfat early lambers may be put with the late lambers to reduce their concentrate consumption. That ration is ½lb (227 grams) daily from six or seven weeks before lambing, building up to 1½lb (680 grams) at lambing. Obviously, the later lambers start this steaming-up process after the others, thereby preventing wasteful and excessive feeding.

If possible, ewe hoggs and gimmers (shearlings or two-tooths) should be penned away from the older ewes, who tend to bully them. If the number of enclosures is limited, I would prefer to give priority to these youngsters which are still growing themselves. Hay should be assessed for quality, and fed in improving order, retaining the best for immediate pre-lambing when the ewes' stomach capacity is affected by the unborn lamb.

Innoculations and Health

In-lamb ewes should be vaccinated with a multi-vaccine at least two weeks, but not more than one month, before lambing. This reinforces the autumn vaccination (*see also* Flock Health).

Swayback is a condition in which new-born lambs partially lose the use of their hind legs. It is the result of copper deficiency, and once there, treatment is too late. Prevention is fortunately now possible. Where swayback is a problem, ewes must be given a copper injection in mid-pregnancy. The vet will advise, and supply the necessary tubes, which are marked to show the requisite amount per sheep. Copper injections are given under the skin, preferably in the neck, and are liable to leave a lump or swelling. Even if this happens, the copper is doing its job, but I find that this 'jag' needs much care.

Feet should be checked eight weeks before lambing. Handling must be done with extreme care at this stage and, once the feet are trimmed, it is better to keep them sound with occasional walks through a footbath. A pre-lambing crutching in lowland flocks keeps the ewes clean for lambing, and udder and rear end clean for suckling lambs.

Scanning

A recent development introduced from Australia is the use of the scanner to determine how many lambs the ewe is carrying. Geld ewes may be separated, and those bearing twins or singles distinguished. Assessing triplets is less straightforward, and 100 per cent accuracy not yet guaranteed. The scanner is a contractor's machine, and its use

The scanner at work, detecting the number of lambs carried.

depends upon availability and price per ewe. In 1984, costs of 50p to £1 per ewe were average, the lower price being for large flocks.

Economic advantages are that geld ewes can be sold fat in the winter high-price period, and do not eat expensive concentrates till their barren condition becomes apparent some weeks later, while single-bearing ewes can be fed less than those carrying twins or triplets. Our knowledge of the nutritional needs of highly productive sheep is sketchy, but a ewe bearing triplets needs 25 per cent more feed than one carrying a single.

PREPARING FOR LAMBING

A fortnight before the first lamb is expected, plans should be put in hand to cope with likely eventualities. If the sheep are to lamb indoors, they must be split into pens of forty as maximum, but preferably pens of twenty.

Whether lambing indoors or outdoors, single pens are needed for new families to be properly mothered up. Hurdles or straw bales suffice, but there should be one pen per six or seven sheep. A pen 5ft × 5ft (1.5m × 1.5m) is ideal; two bales long by one bale wide is often used in practice. There must be a water bucket per pen; newly lambed ewes have great thirsts. Whatever materials are used, the pen sides must be so constructed that lambs cannot hang or smother themselves.

Equipment

Essential aids to be amassed two weeks in advance include:

Clean towels
Carbolic soap
Clean plastic pail
Surgical gloves
Obstetric gel
Disinfectant
Lambing cords
Penicillin
Selection of one inch disposable hypodermic needles (19 g) and

disposable syringes
Glucose
Calcium borogluconate/magnesium sulphate solution
Terramycin spray
Iodine (for navel dressings) – terramycin spray serves equally well,
    but is more expensive
Colostrum in freezer – ewes' or goats', but *not* cows' (may be kept
    from year to year)
Source of hot water
Source of heat
Ewe milk substitute
Needle and suture
Bearing retainer
Bottle teats
Stomach tube
Marking aerosol
Rubber rings and applicator

It is a great benefit if these necessities can be kept together in one
cupboard. They should have their own places so that shortage of any is
easily apparent. Your vet's phone number should be near the phone.

Course

The Agricultural Training Board's lambing courses are the best possible
reminders for both novice and advanced shepherds. Enquire through
your area ATB office, or the NFU.

SIGNS OF LAMBING

Average gestation period of a ewe is 147 days, but marked individual
differences occur. In a small flock it is easy to note when each ewe was
tupped, but lambing will by no means necessarily follow in sequence.

When lambing becomes imminent, the ewe tends to wander away to
seek isolation at the edge of the field or pen. Her udder and teats are full,
she becomes restless and uneasy, and paws or scrapes the ground with
her front feet as if making a nest.

Lambing

Lambing pen essentials: food and water for the ewe, comfort, easy access, freedom from draughts and obstructions on which a lamb might hang or smother.

When lambing begins, she stretches and strains her neck when lying down, typically with the head pointing skywards. The water bladder may appear, or it may have burst and left signs of fluid and moisture. Prolapse of the vagina sometimes occurs at or before lambing, and is countered by insertion of a retainer, to be removed at lambing.

Presentation

In normal presentation, the forelegs appear together, with the nose laid between or on them. Only in the case of a very large lamb is there difficulty with a normal presentation.

Assistance

Birth is a natural process, and help should not be given unless it is needed. Hill shepherds are especially averse to helping shearlings to lamb, as the next time there may be no one on hand to help. Sometimes a hill shepherd delivers an older ewe if dusk is falling and she still has a long way to go, simply to ensure that all is well, although the ewe would doubtless have managed on her own. Such criteria apply less to the small, intensively shepherded flock.

Average duration of lambing is ninety minutes. If there are no results after two hours, or fluids escape with no sign of progress, investigations should be made. If the head appears alone, or the tail appears alone or with one or two hind legs, help must be given. There is risk of infection every time a hand is inserted into the vagina, so hands must be as clean as possible, and nails kept short. A penicillin injection or use of an antiseptic pessary is a worthwhile safeguard for every ewe that has been helped internally.

Examination may show that presentation is normal, but if thirty minutes later there is still no birth, help should be given by traction on the forelegs one by one, then easing the head out by drawing down towards the udder. In most cases this downwards traction towards the udder is the correct one, *not* in line with the spine.

Small hands are an asset, so a woman or child may be of more use than a man. All manipulations must be gentle, and the claws of the foot covered by the fingers to prevent tearing the passage. With twins or triplets, try to identify which limb belongs to which lamb, and start with the lamb nearest the vagina.

Mispresentations

Only the commonest and most straightforward mispresentations are dealt with here. It is better to fall back on experienced or professional help if the more complicated mispresentations occur.

If the head appears with just one front leg, it is usually easier to lamb the ewe in that way, rather than trying to retrieve the backward-laid foreleg. It helps if the ewe is placed on that side away from the retained leg.

When the forelegs are presented correctly, but the head is back, push the lamb gently back in until head and neck can be manipulated into line.

Quick action is needed with any posterior presentations, or the lamb will smother. If two hind legs are presented, make sure they both belong to the same lamb; if only one hind leg, push the lamb back gently between strains, and bring forward the other leg. If there are two hind legs but no tail in sight, try to work it backwards before drawing the lamb.

In a *breech* presentation, only the tail is presented, the hind legs being pointing forward. With an assistant to hold the ewe upright, head downwards and resting on its shoulder, cup the hand over the hind feet, one at a time, and draw them towards the vagina. There is risk of damaging the passage, and it is advisable to take the ewe to a vet if possible. Often it is quicker, and cheaper, to take the ewe to the vet rather than vice versa.

Routine

In some years a number of mispresentations may occur in a small flock, in others scarcely any. Reasons are seldom apparent, and in any event little can be done other than treat in-lamb ewes carefully and gently.

The question of night shepherding arises. In a flock of 40 ewes, most of which held to the first service, only two ewes will lamb each day on average. It won't work out like that; one day there may be three or four, and then two or three days when nothing happens. The most practical advice where indoor lambing is concerned is to have a late look round, say at midnight. If all is well and quiet, it is doubtful whether the loss of sleep and effect on the next day's work is worth a further visit before about 5 a.m. If a ewe is showing obvious signs of lambing, that is a different matter, but it is soul-destroying to get up in the middle of the night for several nights running to no result. The sight and sound of a yardful of Welsh or Swaledale ewes contentedly cudding may be heart-warming, but palls at 3 a.m!

Outdoor Lambing

If lambing is taking place in a field, there is seldom any point in looking round after dark. I have trailed out with a powerful torch, and done nothing more than disturb otherwise peaceful sheep. Outdoor lambing is preferable to disorganised indoor lambing.

One Durham farmer brought 220 Mule ewes in at nights under a very

A well-managed autumn lambing flock can achieve 150 per cent lambs reared, and is attractive to the smallholder as the ewes may be kept on little ground in summer.

large openspan former silage barn and foldyard. 'I went out in the mornings and picked up armfuls of lambs, without a clue as to which belonged to which ewe!' he said. He soon realised that without proper penning and drafting, his flock would lamb better in a four acre paddock, and they did.

AFTER THE BIRTH

After the ewe has lambed, catch her and check that she is milking from both quarters. A gentle squeezing will bring either a drop or a squirt of milk, but don't try this *before* lambing, as the seal is broken and there is risk of infection. Check externally that the ewe has had all her lambs. A dead one left inside may soon kill the ewe also.

If possible check that the ewe has cleansed; a 'hung' cleansing is dangerous, and a vet's job. The lambs should be dressed on their navels with iodine, handiest if in a narrow jar of about one inch diameter. The normal iodine bottle is fiddly when dealing with a bloody navel. Terramycin spray is equally effective, the principle being to prevent infection through the raw navel. I sometimes dress navels twice, once immediately after birth and again a few hours later when they have partially dried and shrivelled.

Ewe and offspring should be placed in a single pen. Ensure that the ewe takes to both her offspring, if twins, and arrange to mother a triplet elsewhere if feasible.

Hypothermia

Chilling or hypothermia means a below-normal body temperature which will soon lead to coma and death if not countered. It is a condition

Numbering lambs with aerosol spray saves many mix-ups.

more often encountered on hill or field than among closely herded sheep, but the smallholder must know what to do to correct it.

The first step is to dry the lamb on a rough sack or towel. Then take it to a source of heat, whether kitchen fire or specially designed lamb warmer. The Moredun Research Institute, 408 Gilmerton Road, Edinburgh, provides details of the latter.

When drying a wet lamb, don't forget that the limbs and tail total a very large surface area. Dry them as well as the torso. Do not put wet lambs into the warmer.

The lamb should be fed after warming, not before. If the lamb is newly-born and its dam's milk is not available, stored colostrum should be used. Lambs over five hours old have very low blood glucose levels, and may have glucose injected into the abdomen; a professional demonstration is needed for this.

For most lambs, a feed by stomach tube immediately after warming is beneficial. The Scottish Colleges recommend use of the stomach tube whenever a newborn lamb has to be fed by the shepherd. A weak lamb sucking from a bottle may inhale milk into its lungs, setting up pneumonia. Also, tube-fed lambs do not become 'bottle-orientated' and refuse the ewe. One Blue-faced Leicester lamb so preferred the bottle that its owner had to milk the ewe twice daily for twelve weeks, and feed the offspring by bottle!

STOMACH TUBE

The stomach tube consists of a clean tube attached to a 50ml syringe. It should be washed after each feed, and sterilised at least once a day. Feeding procedure is:

1. Sit comfortably on a stool or straw bale with the lamb on your lap.
2. Gently introduce the tube (with no syringe attached) into the gullet via the side of the mouth. No force is required. In a large lamb all but 2–5cm (¾–2 inches) of the tube can be easily introduced. If the lamb shows signs of distress, withdraw and start again.
3. Note the lamb's reactions. If the tube is in the stomach, the lamb will show no discomfort and may chew the tube.
4. Attach a syringe of colostrum to the end of the tube. Emit contents slowly over about twenty seconds. Remove empty syringe and attach

Lambing

a full one, and repeat as necessary.
5. Remove syringe and tube as a single unit.
6. Wash and disinfect tube and syringe.

Awkward Cases

One well-known Borders farm manager marked one set of twins AC with colour spray on their flanks. Asked the reason for such complicated lettering, he explained; 'Awkward Customers'! Every shepherd experiences such cases, for lambs often seem to have an inborn desire to die, and the shepherd's job is to thwart that wish till they are properly on their feet and enjoying life. Remember that the most pitiful, uncooperative newly-born lamb may be quite indistinguishable from the rest at sale time.

Occasionally a ewe, often a shearling, turns against her only offspring. She must be restrained, either in a lamb adopter or tied by the neck or horns. The lamb is held in the appropriate position and a little milk squirted into its mouth to start it feeding. Once a ewe's own milk has passed through the lamb, she is generally more inclined to accept it. Recognition is largely a matter of smell.

Twins

If a ewe refuses one of a pair, the lamb adopter is again brought into play. Failing this, try tying up the ewe and removing the favoured lamb for a time. As milk accumulates in the udder, supervise the reject's suckling, and persevere till both are accepted. This may take days; we had a Blue-faced Leicester that took six weeks to accept one tup lamb of a pair. Eventually it did as well as the other.

Mothering On

A ewe may lose her own lamb, or have a single and sufficient milk for twins. This is the time to put on any spare lamb, either a triplet, a twin whose mother is milking on only one side, or an orphan. The ideal time is when the foster mother is actually lambing.

If you think she is only having a single, be ready with the orphan, burst the water bag over it, rub it well in the fluids and with the ewe's

56

The creep feed in the corner enables these fine Suffolk lambs to feed away from their dams.

own lamb, and remove the latter. Give the ewe the orphan to lick, and if all is well re-introduce her own lamb after fifteen minutes. This theory is upset if the ewe then turns against her own lamb, and the shepherd's best-laid schemes truly gang agley if the ewe produces a second lamb. But it is worth a try.

Recent research has put hessian at the top of the list as a material to aid mothering-on. It is less messy than the traditional skinning of the dead lamb, and fixing the skin on the orphan. A variety of strong scents may be used to drown the difference in smell, but some ewes persistently refuse whatever the method.

If the shepherd turns out a milking ewe without a lamb, he has failed. The object with an upland flock is strong singles and as many twins as possible, and twins all round in a lowland flock.

CASTRATION

Uncastrated males normally grow and convert slightly better than castrates. In a flock producing fat lambs straight off the dam, it may be

A simple lamb feeding device.

unnecessary to castrate. The snag here is that there are usually some lambs that have to be run on to a later stage, and if they are entire they become a nuisance, cannot be run with females at certain seasons, and may show 'tuppy' characteristics that turn the grader against them.

Castration is thus generally the soundest policy, except of course where rams are being retained for breeding. In their case, more are left uncastrated than are needed as studs, and weeding out takes place throughout growth until only the desired number remains.

Method

Rubber rings are simplest. An 'elastrator' must be bought or borrowed, its function being to open the ring so that it can slide over the testicles. The operation is simpler with an assistant to hold the lamb, laid on its

A Suffolk cross receives its identity mark.

Forward creep grazing, with the lambs on fresh grass ahead of their dams.

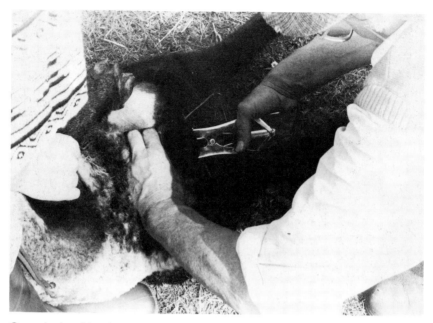

Castration by rubber ring.

back between the knees. The first step is to locate *both* testicles; never put a ring over just one because you can't find the other, or you are left with a 'rig', as much a nuisance as an entire.

Both testicles are eased into the scrotal sac, the ring slipped over and released. Be sure that it clears the rudimentary teats. Lambs show some discomfort for about fifteen minutes, after which there should be no further trouble. But if that period is prolonged, cut off the ring and leave for a few days.

Many competent shepherds castrate at twenty-four hours, or as soon as the ewe and lambs are turned out with a group. That is all right when all is well, but I prefer to wait another day or two if there is the slightest doubt whether the lamb is thriving.

Tailing is done at the same time, and the common practice when dealing with a 'rig' is to leave his tail long if other lambs are tailed. If all tails are left long, put a clear distinguishing mark on him.

Very short tailing is now illegal. Enough stump must be left to cover the anus or vagina, according to sex. Check the number of your fingers needed to do this, and grasp the tail root accordingly. If the elastrator is put on with the points forward, the ring is easily left in the desired

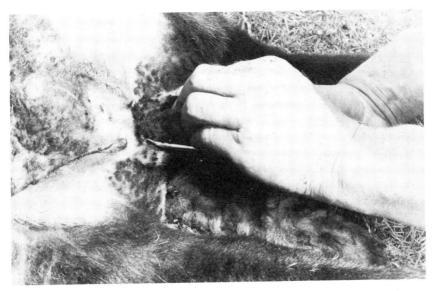

It is essential to clear the rudimentary teats.

Ring in place, and pliers ready for detaching.

position, and the gadget slid away.

Unless rings are applied during the first week of life, they must be accompanied by local anaesthetic. It is an offence to castrate male sheep by any method without an anaesthetic after three months of age, and only a qualified vet may perform the operation on a ram that has reached the age of six months.

MARKING AND RECORDING

The identity of each new lamb should be recorded. Sheep are traditionally marked in three ways, or a combination of them. The fleece mark is of varying colour, and is generally the flock mark. To the flock mark may be added another denoting a particular heft or cut of sheep. Green, black, blue and red are the chief colours. Do not choose the rump, as that space is needed for the tup's raddle. A system of ear notching is also used, usually on hill flocks, but sometimes for pedigree lowland sheep. Combinations of notches at varying points on either ear allow a wide range. The third method is horn burning, usually with the owner's initials, or his predecessor's. To these are added tags, very handy as numbers may be added to the owner's initials or name. There is a choice of systems; some shepherds use a different coloured tag for each ram, others a different one for each year.

At Redesdale Experimental Husbandry Farm, Otterburn, Northumberland, Harry Jameson herds the Dargues Hope unit. He uses coloured tags, one for each age of sheep which are normally drafted at five or six-shear.

All male lambs receive an orange tag, but potential stock rams are double-tagged. Ewe lambs are tagged according to their sire, using numbers in sequence. One ram's offspring are allocated tags one to 50, the next 51 to 100, and so on. At any time in its life, any lamb can be identified without reference to files.

At lambing, the shepherd carries a hard-backed notebook, and on tagging a new-born lamb enters the tag number, sex, date of birth and dam's number. If a lamb is fostered onto another ewe, it is essential to note the change at the time. Transfers are made from the notebook to a permanent record daily, showing the same details.

5   Shearing and Dipping

VALUE OF WOOL

Wool was once the most valuable commodity per pound that the farmer produced. Today that is unfortunately not the case, and the returns from wool compare poorly with fat lamb. A draft hill ewe may produce lambs worth £60 or more, but less than £2 worth of wool.

Other breeds, notably Devons and other longwools, provide sufficient fleece to really count. In any event, the wool is there as a by-product, so the best use should be made of it.

Presentation

'Deplorable' was how one experienced grader described the average presentation of wool. Farmers lost money largely through carelessness; badly wrapped, dirty fleeces showing lack of pride, and no interest beyond finishing the job.

Shearing takes place mainly in summer. Winter shearing of housed ewes is becoming more popular, but in either case the preparations for shearing are identical.

Fleeces must be dry at shearing, so provision of clean, covered accommodation to house the sheep overnight is an advantage. Slatted floors are ideal, but failing that, clean green crop can be used. Docks and nettles growing abundantly at that season are surprisingly good, but sawdust or straw bedding are quite unsuitable, and will contaminate the fleece. An advantage of overnight housing is that stomachs are comparatively empty, which is safer and more comfortable than working on a full sheep.

If a contractor is employed, care must be taken not to waste his time. A small catching pen should be provided, to take about ten sheep, and preferably fitted with a sprung, self-closing door. A wooden platform is best for the actual shearing, though a sheet may be used, or the well-tried method of squares of clean turf.

The smallholder may prefer to shear his own sheep, but a machine with shearing head, combs and cutters is now so expensive that the economics must be carefully weighed. Hand shearing is still practised, even for quite large flocks, but is hard work, and there is difficulty in buying strong shears that keep their edge. When shearing a small number oneself, the task is finished just as one is working up speed!

Owners of more than four sheep must sell all the fleeces through the British Wool Marketing Board, who provide wool sheets, waxed twine for sewing, and labels. The sheet should be suspended near the wrapping table. Besides the shearing equipment, wound dressing in cream or spray form should be on hand for the inevitable cuts. Each sheep must be marked after shearing, so the paint can and brand are needed.

Sheep are best shorn when the wool 'rises' or starts to part from the skin, attached by fibres distinct from the mass of fleece. To shear at other times makes for harder work. 'Wean early and shear late' is a proven maxim, and to shear in May can result in losses, apart from the cruelty of depriving animals of their natural protection and having to turn them out in driving rain. But in the south, late May shearing is favoured, as it ends the hazard of sheep becoming cast or mislaid.

SHEARING

Machines are suspended so that the drop cable end almost touches the floor. A sheep is drawn from the catching pen, either by catching above the hock and dragging out backwards, or straddling it. A horned sheep is easier, but hold close to the head, or with an older sheep the horn may break off.

The sheep is then set on its rump. A small sheep may be lifted, but a large one is rolled over by pushing a hand into the lisk or top of thigh, and the other under the neck, and setting the sheep upright. A right-handed shearer then pushes the neck down on the left, in which position the animal will sit quietly. By far the best way to learn to shear is to attend the excellent classes run by the Wool Board. Failing that, remember the basics:

1. Minimum physical effort – no flourishes.
2. Control of the sheep – sheep moving into position for the next blow all the time, and balanced.
3. Minimum number of blows – fill the comb so that in most places you use the full width and not just half. Try to use exactly the same routine and virtually the same number of blows on every sheep. But difficult bits round the crutch, under the tail may take more on some than others.
4. No double cutting – this is wasteful of time and wool.
5. Use of the left hand – constantly preparing the way for the hand-piece. Only practice can bring that safe, close co-ordination; watch a top shearer's left hand.
6. Don't fight the sheep – there's always another coming, and perhaps a hundred after. As champion shearer Godfrey Bowen said, it is like a wrestler taking on a fresh man every round of a 200 round fight!

Method

Godfrey Bowen's order of shearing is to start with the brisket, where the wool is padded hard. Continue on the belly, working across, keeping the skin taut with the left hand. Special care is needed round the male's pizzle and the female's udder and teats. Then, leaning the sheep over, shear the tail wool.

The left hind leg is shorn, the leg being straightened by thrusting the clenched left fist into the flank, an impressive move to non-shearers. This tightens the skin, the objective in every part.

The sheep is then set upright again, and the neck, head and left foreleg shorn. That prepares the way for the all-important 'long blows' which start where the belly wool was removed, from hind legs to head, with the sheep lying lengthways and being gradually rolled the while. These blows are parallel to the spine, which is shorn, and followed by one long blow beyond the spine.

The sheep is now laid on its right side, and the right cheek is lifted and shorn, followed by the neck and shoulder at a 45 degree angle. The sheep is raised the while by gradually stepping back, but keeping the skin taut to avoid nicking. The sheep is pulled right over till it is seated on its left rump, and the hind leg reached. When finishing, pass the left hand over the base of the spine just above the tail, for there is often an unclipped bunch of strands which suffice to trail the entire fleece along as the animal is released.

Wrapping

A raised, slatted table is a boon for wrapping, failing which a clean floor is used. The fleece is taken by the hind legs and flung bodily upwards

Shearing – opening out the neck after shearing the belly and near hind leg.

and outwards, so that it descends flesh side down like a rug, with dust and loose fibres floating away. Rough Fells, Herdwicks and some Blackfaces are wrapped flesh side up.

Any foreign matter, such as grass, straw or twigs, is picked off, as are any daggs that have been missed. Flanks are flipped towards the centre making a parallelogram about two feet wide. The britch is then turned in and rolled firmly and neatly towards the neck.

The neck wool is drawn out to form a band, long enough to encircle the wrapped fleece, but not too tight. This band is tightened round the rolled fleece and tucked under itself. The job must be firm enough to withstand several handlings.

Starting on the head after the sheep's near or left side has been shorn.

Shearing and Dipping

FAULTS

Faults include bloom dip, discouraged by the Wool Board, but used on some breeding stock supposedly to enhance appearances. Such tinted wool has a limited use, and is penalised. Tar, pitch, oil paint and creosote are almost things of the past. Binder or baler twine should not

Continuing from the head diagonally across the neck, foreleg, and down the flank.

be allowed near the sheep or sheets. A few strands can completely ruin a fleece, being almost impossible to spot before processing. A few polythene fibres in a soft vest or pullover are just what the wearer does not want.

Wet Wool

On no account allow a contractor to start before fleeces are dry. Shearing damp sheep is a quite serious health hazard and, if a fleece has for some reason been clipped damp, it must be hung up to dry before rolling and packing.

Daggings and loose wool are packed separately. Solid daggings are probably best used by soaking for the garden, but in any event should never be packed in polythene sacks, which do not allow the wool to breathe, and rotting soon occurs. Hessian sacks are best.

Rolled fleeces and bagged daggs are packed into suspended wool sheets. Be sure to pack tight into the corners, and systematically thereafter, otherwise a sloppy, unwieldy mass will result. When full, the sheets are stitched with waxed twine, and labelled. They should then be stacked on timbers, never on brick or concrete, or they will sweat.

Mothering Up

A vital aspect of clipping days is mothering-up afterwards. A proportion of lambs may not recognise their sheared dam, and run madly from one ewe to another. The shepherd *must* see that each and every lamb is happily coupled up before night falls, and in a medium sized flock this can be achieved. In large flocks there tend to be a few which do not mother up, especially if a long walk back to the fell is entailed.

Winter Wool

More and more housed ewes are being winter-clipped, and farmers ask whether the clip can be as valuable as in the normal season. It can, the main difference being felt the first time, when a fleece has had time to grow only from July to January. In following seasons it has the full twelve months. The Board will both receive and pay for winter wool as soon as it is clipped.

Shearing competition; a spectator sport in New Zealand, and gaining popularity in Britain.

DIPPING

Sheep are heir to a number of skin parasites, the most dangerous being sheep scab. Dipping would be essential to successful sheep husbandry even if sheep scab were eradicated, and serves several functions. Ticks, keds and lice can all affect health and diminish performance.

The *winter dip* includes waterproofing ingredients which help the fleece shed water. Before the advent of modern dips, sheep were *salved* by a mixture of tar and butter, both bought by the barrel, the fleece being parted in strips and laboriously worked in by hand. Ernest Metcalfe of Rosedale, North Yorkshire, is one of the few survivors of those who actually carried out this work.

Then farmers went to the other extreme, and judged dipping by the number they could push through in an hour. A more commonsense approach now prevails; the whole operation is an expensive waste of time if not done thoroughly, which usually means immersion for a

These Mule lambs have been winter shorn, and may be more densely packed in the pens.

minute.

That period must be timed accurately. Apart from the compulsory dips, notified to local authority or agricultural advisory service, there is an increasing trend towards more frequent dipping. With an efficient set-up it does not take long to put the whole flock through once a month from May to September, and only on dates that fall in the compulsory period will the full minute be required. For the rest, a monthly half-minute immersion is more effective than merely complying with the legal minimum requirements.

Principles of Dipping

Sheep must *not* be dipped:

1. On over-full stomachs, or when hot and weary.
2. In rain, or if rain threatens before fleeces can dry.

3. Too late in the evening, so that sheep remain wet all night.
4. With open cuts or wounds. Keep the affected part clear of dip.
5. Immediately after shearing, as there is too little wool to retain the dip, and sun scald is a possibility. Ten to fourteen days after shearing is ideal.

Summer dipping is mainly against the blow fly. In pre-1939 flocks, summer could be a nightmare, with the shepherd on daily alert for the tell-tale self-biting that indicated fly 'strike', the maggots eating into living flesh.

Then dieldrin was introduced, and put an end to all that. It was banned because its own persistence was proving damaging to other stock, and eventually to man. The latest dips do not compare with dieldrin, but remain effective for at least 56 days.

A circular swim bath enables sheep to be submerged for as long as necessary without holding them back manually.

Show Dips

Proprietary dips are used before show and sale. They are particularly important in breeds like the Blue-faced Leicester, where a curl in the wool is desirable. These dips are usually extra-strong, so do use precisely as instructed on the label.

In any dipping system, a stepped ramp is provided for the sheep to clamber out.

Shearing and Dipping

Bloom dips colour the wool, to make the sheep more attractive to buyers. Bloom dipping is frowned on by the Wool Board as the wool is worth so little, but as long as the live animal's value is many times that of a fleece that will be sold by someone else anyway, the practice may well continue until such dips are barred by law.

A 'plunger' is needed to dip heads.

Process

To start the season, the dipper must be filled to the appropriate level with *clean* water. A new bath must have levels marked accurately. On a purchased bath, these should be shown, but a farm-built bath must have its levels marked by using a container of known capacity; a ten-gallon milk churn is very handy. Above about 50 gallons, draw an indelible mark to show every ten gallons. This is essential for accurate replenishing.

Replenishing

When ready for dipping, add the dip to the clean water at the prescribed rate, usually one part to two hundred. Sheep soon lower the level, even with an efficient drainage system, to the point where it is less easy to submerge the sheep. The dip must then be topped up and, because dip is filtered out by the wool, a stronger solution is needed. This is often one part to eighty, but directions are clear.

Method

Gentle handling is the watchword. Even ewes quite heavy in-lamb may be safely dipped if care is taken, and in most systems the animal is lowered in backwards. A T-shaped ducking stick is used to push the head under two or three times.

If the water is very cold, start with a few strong sheep. Their body heat soon takes off the chill that may affect old-stagers. A breezy, sunny day with slight cloud cover is ideal, but in a wet season one has to dip when one can, but not when actually raining or about to rain.

Discard old, soiled dip. It is a waste of money to add expensive fresh dip to stale, mucky sludge. A rule of thumb is to dip no more sheep than three times the bath's gallon capacity before starting again.

6 Equipment

It has been said that the only equipment a hill shepherd needs is a pair of strong boots and a crook. That was before the days of vaccines and penicillin, but the basic truth still holds; shepherding is a profession far more reliant on basic skills than on gadgetry.

However, a number of devices have been invented to ease the shepherd's lot. They include restrainers for examining and working on the sheep, lambing aids, and shearing machines. First of all, though, the flock must be kept where it belongs.

FENCES

The first essential is a sheep-proof boundary fence. Then, if the sheep break out, they are troubling you and not your neighbour. 'Good fences make good neighbours' is a well-tried country saying, and neighbourliness is an essential part of country life.

Sheep are most apt to stray in spring, when there is a bite of new grass, and they will make far bolder attempts to break out then than in high summer. Three main types of fence will be considered; mesh, strained plain wire and electrified wire.

Pig Netting

Pig netting is a trade name for galvanised, strong wire mesh, usually sold in 55 yard (50 metre) lengths. When topped with one or two plain or barbed wires, it makes a fence that will contain any sheep in normal circumstances. It requires a straining post at either end of each net, at corners and at changes in direction. The following list of tools and equipment is the minimum needed to set a mesh fence:

Equipment

Spade
Crowbar
Sledge hammer
Hand hammer and saw
Wire strainers
Billhook and scythe if a path needs clearing
Nails
Staples
Pliers or device for joining wires
Nets
Plain and/or barbed wire
Straining posts
Leaners and struts
Intermediate stakes

The first step is to mark out the fence line. This should be as straight as possible, which may mean following an existing line, or striking a new one if the old one can be improved upon. In the case of a boundary or march fence, ensure that your line is correct, and that your neighbour and his landlord, and your landlord agree. Wrongly sited fences can cause no end of future trouble.

The materials are then set out, with heavy straining posts placed as near their positions as possible, stakes thrown in a line every three or four yards, and barbed and plain wire at one end.

STARTING THE WORK

Straining posts are set first. They should be at least 8ft (2.4m) long and of 7in (18cm) diameter, or square. They are set 2ft 6in to 3ft (76–90cm) into the ground, and the hole should be dug no wider than necessary to work. The 'front' of the post (the side from which the wire is being strained) should be set against solid earth, and five degrees away from the direction of the strain, for it will ease forward into the perpendicular.

A hefty stone or clog of hardwood is set against the base of the 'heel', to keep it in place, and another just below ground level on the face side.

Filling starts, using dry, sound material rammed home really hard from the base. If the first foot is not solid, the post will wave about uselessly. The rest is replaced a bit at a time, and rammed home with a handy stake or beater.

The next straining post is then set, and the two joined by a wire to show the line of fence, *before* the struts are applied. This shows where the struts should be set. They should be 7 or 8ft (2.1–2.4m) long; if much shorter, they make too steep an angle, a pivot around which the post will raise itself slowly and remorselessly from the ground during straining, not a pretty sight or one for delicate ears.

SETTING THE STRUT

The strut is leaned against the post along the fence line, some 2ft 6in (76cm) above ground level. A neat notch is cut into the post face, to take the head of the strut exactly. Leave it in position, and mark where it rests on the ground with a spade. Then dig *straight* down. Turf and soil are removed, to 1ft (30cm) depth, and a flat stone set to take the end of the strut. The strut is then set against this stone, and the head jammed into the nick on the post. Secure there with two 4in (10cm) nails.

SETTING THE NET

The net is now run out, and the far end weighted down, otherwise it will rejoin you! One end is fixed to the first straining post by taking it completely round the post, starting about 4in (10cm) above ground level, with each loose end twisted on the corresponding net section. Staple it securely to the post, two nails for top and bottom, and one on each mesh elsewhere, all driven home. These are the only staples to be driven right home; others support the wire but allow tightening through them.

A monkey wrench is the easiest means of tightening. Do not over-tighten. Galvanizing is cracked through undue pressure, and that starts the end of the net's life.

INTERMEDIATES

Not till the net is tightened are the intermediate stakes set unless a tractor-powered post driver is used. Use a rail cut the exact length between them, which should not be more than three and a half yards

(3.2m). Prick a hole with the crowbar in the appropriate place, but only as deep as necessary in the conditions.

Drive in each stake with a sledge hammer till firm. Staple alternate meshes to stakes, but do *not* drive staples home. Tap in any uneven stakes, and put on the top wire. Finally, make up any low places with sods, or rails if necessary, and clear up all bits and pieces. Leave everything tidy.

Plain Wire Fence

High tensile wire may be strained over much greater distances, but is more applicable to large fields.

Summary

With either mesh or strained plain wire fencing, the order of work is:

1. Lay out materials
2. Set straining posts
3. Run out wire or net
4. Slightly tighten, to give line
5. Set struts
6. Tighten wire or net
7. Set intermediate stakes
8. Staple net to them
9. Set top wire
10. Make up low places
11. Tidy surplus materials

Electrified Fencing

Mains electric fencing is for wide stretches of hill land rather than for smallholdings. It has applications for temporary fencing, however, and for folding over either grass or arable crops.

The Ridley Rappa uses High Powered Energisers run off either mains or 12 volt wet battery. It can be moved constantly about the farm, so is not lying idle for most of the year. Among its uses are:

Quite sheep-proof. Rylock high tensile stock fence requiring fewer straining posts. Note wire struts holding down the fence in lower places, and the tight barb on top.

Making temporary paddocks at lambing time.

Making newly-occupied land workable, giving time to erect permanent fences.

Folding swedes or sugar beet tops.

Allowing acreage to be extended through renting badly-fenced grass parks at a much lower rent than well-fenced fields.

As guard against corn crops in a ley system.

Up to four wires are carried by the Ridley Rappa on a wheelbarrow type frame. The gearing allows sufficient tension to prevent entanglement, either when paying out the wires or reeling them in, and the tension is easily adjusted.

Light intermediate posts are carried on the machine, and surplus wire remains on the reels which are then fitted to the reel posts. A basic set for 400 metres of fence consists of one reel post, one anchor post, metal stakes, insulators and reels carrying the wire. For sheep, three wires are recommended, and the system can also be used against rabbits.

An electric fencing system at the National Agricultural Centre, Stoneleigh, Warwickshire.

The Ridley Rappa – a quick method of electric fencing.

Section through a drystone wall, showing method of construction. Note the throughbands halfway up, the heavy base stones and the coping on top.

Wall in course of erection, using guide lines.

82

This wall was built from stones off reclaimed land on either side, and incorporates big rocks without moving them.

The art of hedge laying is well worth acquiring.

Equipment

HOUSING

One great benefit of sheep is that they do not need large capital expenditure on housing. For generations, most British flocks were wintered in fields, and are still perfectly capable of doing so.

Housing the sheep has greater benefits for the land and the shepherd than for the sheep. A grass paddled by feeding ewes all winter starts growth many days or even weeks later than a rested pasture, and that is a strong reason for the smallholder to consider housing. Such a rest helps control parasitic worms, although it does not control nematodirus, eggs of which are shed in spring and autumn.

Shepherding is easier and pleasanter. Winter ruts in fields are obviated, and no tractor time is involved for feeding. Lambing indoors is much more certain than braving an English spring outside, and this saving alone may justify housing. The flock can be inspected during the night if necessary.

Ventilation

The main principle to bear in mind is that the temperature inside the house should be similar to that outside. A warm, enclosed shed is quite unsuitable for sheep, for pneumonia is a real threat. The sheep must be free from draughts, which means protective walling to about 4ft (1.2m). They must have air circulating above them, helped by ample ridge ventilation.

Structures

Regarding structure, this varies from pole barns made out of home-sawn timbers to quite elaborate concrete-and-asbestos custom-made sheds. The cost of the latter may be £30 a ewe housed after grant, the former £10 or even less.

Depending on site, some protection may be needed above the 4ft (1.2m) wall, varying from spaced boards to Netlon. Open space may be feasible on the lee side. Snow swirling in can be a problem; as it falls, it does not bother the sheep, but if it remains and melts it causes soggy bedding. The roof should have ridge or apex ventilation, nothing more elaborate than a space between the two pitches.

Concreting the sheep house floor is doubtful economics. A surface that allows water to seep away while being stable enough for mechanical cleaning suffices.

SLATTED FLOORS

Slatted floors save bedding and keep hoofs healthier. An area of 15 square feet (1.4 square metres) for large ewes on straw is advocated. For smaller or shorn ewes and slatted floors, less is feasible. Even more important is feeding space; a minimum of 18in (46cm) per ewe should be allowed. Bear in mind the need for extra emergency space if bad weather prevents 'turnout' in spring.

Unlike cows, sheep do not feed in turns. They all come to the trough together, and any that are pushed out tend to not bother, and so are weaker the next time. While 18in may seem generous at the start, it will

Blackfaced ewe lambs on slatted floors.

not be so when the ewes are heavy in-lamb.

OUTDOOR SLATS

Another possibility is outdoor slats. A sheltered part of the homestead is chosen, and sheep do winter well on this system in areas of moderate rainfall. I have not seen its application in high rainfall areas. Hay troughs should be covered. By buying secondhand slats, housing as cheap as £2 per ewe can be provided.

Water

Water must be supplied at all times. Hill lambs seem to prefer running water, so a spring or stream diverted through the sheep house is ideal. Milking ewes drink a colossal amount of water compared with in-lambers. Care must be taken in design to ensure that the supply does not freeze.

Northumberland sheep farmer Bill Richardson in his slatted-floored shearing shed.

Pen Divisions

Pen divisions must be fool-proof, and high enough to dissuade heavily in-lamb ewes from attempting to scramble over them. Troughs form the best divisions, thereby saving space. No pen should hold more than forty ewes; twenty is better, and one great benefit of housing is that the various classes and ages can be separated. This is far more important than any time saved in foddering.

Though the main flock may be of full-mouthed, mature ewes, these should be handled at least monthly during pregnancy, and any lean ones drawn off into a separate group and given extra feed. Hoggs or tegs merit a pen to themselves, as do shearlings in-lamb for the first time. It is no bad thing to pen two-shears on their own also; at that stage they are coming to maximum value, and care now builds future reserves. Any

Haybox used at Redesdale Experimental Husbandry Farm, Otterburn, Northumberland.

The Universal Sheep Cradle for easy working, from Universal Livestock Services, Banbury.

Wide entry footbath in which sheep can be held. It can be sited at pen or fields exits.

broken-mouthed ewes can join the lean brigade if necessary.

Housing enables ewes to be kept to a greater age. In Iceland, where all sheep are winter-housed, there are pens of 'teenagers' in every shed, whereas in Britain we regard anything above seven years old as noteworthy.

Troughs

Troughs may be designed for feeding, as pen barriers and inspection platforms. It is a great benefit to be able to walk above the sheep rather than amongst them. Departures from the norm, sought by every stockman on every inspection, are easier to spot. Time is saved, and feeding is pleasanter for man and beast.

Useful dimensions for troughs are 24in (61cm) wide, with sides 9in (23cm) high, standing on 8in (20cm) legs, and with rails 11 and 7in (28 and 18cm) above the sides.

Doors

Doors need only be small for taking in fodder, but a larger door or wall section should be available for cleaning out by tractor. With straw bedding, consider the fire risk, and have emergency exits if necessary.

HANDLING YARDS

The handling pens should be sited with convenient access from where the sheep will be gathered. Water, preferably piped, should be available, and a means of emptying the bath by drain to a soakaway, which entails a fall of at least one in 50 from bottom of bath to top of soakaway.

A gently sloping site eases drainage, and sheep move more readily uphill. Some shelter is desirable, according to district, and working and dipping layouts partly or wholly under cover are becoming more common.

A great many ingenious devices have been incorporated in handling systems. Dipping is compulsory, and it is well worth installing a workmanlike dipping arrangement. There is no point in making the bath too big. A 200 gallon (900 litre) bath will suffice for medium sized flocks;

those time-saving 600 gallon (2700 litre) designs take a great deal of expensive dip, of limited life.

While the dip bath is the focal point, other essential components are:

1. Receiving yards, which take the whole flock as it arrives from the field.
2. Forward yards or pens, much smaller and taking a small mob of sheep, leading to;
3. Crush or forcing pens, usually one or two long, narrow pens not more than two sheep wide according to breed. Sheep are dosed or vaccinated in them.
4. Drafting race terminating in a drafting gate or gates. This race is long and narrow and wide enough for only one sheep. At its exit, sheep are diverted two or more ways, ewes from lambs, or males from females, after which they enter small drafting pens. A four-way drafting or shedding gate is feasible, and on Australian sheep stations its operation is given to the newcomer from England, who becomes a ready butt for Outback witticisms if an earmark is wrongly deciphered.
5. A footbath is essential, and may be incorporated in one of the races, and covered when not in use.
6. Draining pens slope back towards the bath. They allow newly dipped sheep to stand, and excess dip to drop from their fleeces and return to the bath.
7. Holding pens take sheep after being dipped and before returning to pasture. They should correspond in size with the receiving pens.

Concrete floors are essential for the catching pens and the drainage pen, and desirable elsewhere in the yards. However, there is a strong case for using hardcore initially, and concrete only after experience has shown the design to be correct.

7 Flock Health

RANGE OF DISEASES

To itemise every disease to which sheep are heir would baffle the novice, or even dissuade him from embarking on the shepherding craft. Fortunately, thanks to the often unsung dedication of agricultural scientists, many diseases which were a nightmare to our grandparents are now controlled by vaccines. It is quite amazing how quickly the new skills of disease prevention have been mastered by working shepherds.

Multiple Vaccines

Multi-vaccines, 8 in one or 7 in one, are essential for sheep health. These are used to prevent clostridial, mainly soil-borne diseases. When buying sheep, ascertain if they are in 'the system', and have had routine vaccinations. If you are at all doubtful, start again from scratch; no harm will be done if the vaccines have already been given.

Each sheep receives the amount stated on the package, usually 2ml, followed by a booster from four to six weeks later. That is vital, otherwise the first injection is ineffective. Thereafter, an injection once or twice a year is sufficient. There is a time lapse of some fourteen days between vaccination and the development of significant levels of immunity.

Lambs whose dams have been properly vaccinated receive protection through the ewe's milk for the first four to six weeks of life. Then they are given a single injection, which suffices till autumn.

The diseases covered are:

1. Pulpy kidney, which attacks thriving lambs. They are found dead; there is no second chance.
2. Braxy, associated with hoar frosts on autumn mornings, and again giving no warning.

3. Blackleg.
4. Lamb dysentry, a heart-breaking condition to which lambs a few days old succumb, after being born perfectly healthy.
5. Tetanus (sheep found dead).
6. Entero toxaemia.
7. Black disease.
8. Bacillary haemoglobinuria.

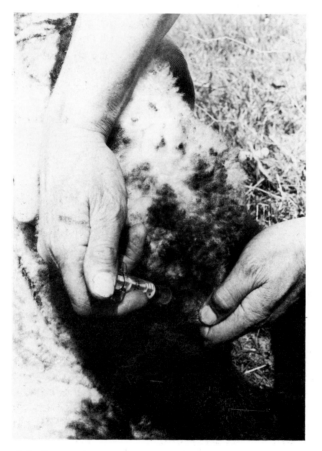

Injections are given carefully, under a fold of neck skin.

METHOD

The smallholder has an economic problem with vaccines. Once opened, the container is liable to infection from needles and the air, and so its contents cannot be stored until the next vaccination or even till the next day. Bacteria multiply rapidly in twelve to twenty-four hours. Most brands of vaccine are for a minimum of fifty sheep per pack, so if only twenty-five sheep are to be injected, the cost per head doubles, unless a neighbour in the same situation is available. Even with a large flock, a part-used bottle is often thrown away.

Incorrect or careless injecting produces abscesses which reduce carcass value. The Meat and Livestock Commission advocate vaccinating in the upper third of the neck. Part the fleece by lifting a fold of skin, and vaccinate just under the skin; do not penetrate the underlying muscle, unless an intramuscular injection such as penicillin is needed, when the rump is commonly used.

Insert the needle parallel to the body and not at a steep angle. Work quietly, smoothly and carefully.

EQUIPMENT

Modern multidose disposable equipment reduces infection risks. Make no attempt to save on needles; a fresh one should be used after every twenty-five or so sheep, and after a break in work. Sterilised needles should be used, handled by the base and not the point; don't 'clean' a dropped one on your lambing trousers! Injections should be done only when the sheep are dry and clean, for when wet or dirty sheep are injected, the needle is contaminated for all subsequent injections.

SYSTEM

Do work out a proper system for vaccinating. Have the sheep under control in a race or small pen. If in a race, and there is no possible chance of breaking back or jumping out, release each sheep as it is jabbed. If in a pen, give each a light colour mark from the keel stick, not from the permanent marker can. Sheep should carry the minimum identification marks, especially if they are to be sold, for buyers do like room to put their own! Add this light mark as each sheep is injected or dosed; it is infuriating to think you can remember the last few, and then find yourself interrupted. Sheep can be turned out of the pen as they are dosed, but the last few are then more difficult to catch.

DOSING

A whole range of dosing guns and equipment is available. There's a lot to be said for the old-fashioned method of bottling; you are then certain just how much a sheep has had, and are not relying on some gadget functioning correctly. For a small flock the extra time amounts to little, especially if voluntary unpaid help is available to fill the bottles!

Stand by the animal's head, keep the head level, insert the bottle or gun into the corner of the mouth, and tip or operate the trigger. Never ram in the nozzle or bottle, and avoid carelessness that may result in fluid going into the lungs.

Stomach Worm Control

The other major health programme, more flexible than vaccinating, is control of stomach worms. 'A sheep's worst enemy is another sheep' refers chiefly to the rapid spread of intestinal worms. Though 'clean grazing', using ground not grazed by sheep in the previous year, is the ideal, it is seldom practical on the small farm. Properly used modern anthelmintics enable heavy sheep stocking to continue.

The more intensive the system, the more dosing will be needed. On an extensive hill, dosing in autumn may suffice. In most grassland systems, the ewes are dosed in autumn before tupping, and again in mid-pregnancy or just before lambing. There is much evidence to support worming ewes six weeks before lambing.

Lambs are dosed as necessary. Once a month is standard on some well-run holdings. Others watch carefully and dose only if dirty tails appear.

Ring the changes on the different brands. Some are more effective against a particular type of round worm, at a particular stage, than others. The real killer is *nematodirus*, which carries over-year, and which lambs pick up from grass. It causes diarrhoea, wasting and death within a few days. Specific anti-nematodirus wormers are now available.

Worms multiply rapidly, especially in hot, humid weather, and the shepherd must be on guard with a close check on every lamb every day during the growing season. That sounds like a lot of work; it isn't really, once confidence is built up, and you can tell a healthy sheep at a glance.

A Rough Fell ram receiving anti-headfly treatment. Only recently has real progress against this scourge been made.

NOTIFIABLE DISEASES

Sheep Scab

Sheep scab has made its unwelcome reappearance. It causes itching so intense that the sheep knows no rest, becomes emaciated and eventually dies. Double dipping between specified dates is now compulsory, but as the Ministry of Agriculture and farmers do not always agree, and changes in dates are made, these should be checked locally. The Ministry, the Scottish colleges or the health authorities are the people concerned.

Scrapie

This most complex and puzzling disease attacks the nervous system, causing severe itching, lack of co-ordination and death. There is no cure, nor prospect of one. All the sheep farmer can do is try and buy scrapie-free stock in the first place, and keep them that way. Some breeds claim to be scrapie-free. Sheep under eighteen months are seldom affected, and if foundation stock of five years old and upwards is bought, chances of an outbreak are lessened but not completely obviated.

Flock recording is well worth the time and expense for several reasons, and the prospect of controlling scrapie outbreaks through selection is an added boost. If scrapie appears in the flock, the only way to try and control it is by culling all descendants and known ancestors of the sheep concerned. This means accurate recording and a willingness to be ruthless when it means disposing of some favourite sheep of desirable type.

It is believed that a gene controls the timing of scrapie, and it may incubate till beyond the time of the sheep's death from other causes. That merely indicates its complexity. Try to avoid scrapie by buying from scrapie-free flocks if you can, in itself no easy matter in view of the secrecy which has surrounded the condition. Turn to the Agricultural Development and Advisory Service, the Scottish colleges or the Meat and Livestock Commission if you run into trouble. Knowledge is being gained all the time, and a number of flocks have successfully eradicated the disease after a bad infection.

PREVENTABLE DISEASES

Swayback

Swayback is a deficiency of copper affecting the brain and muscle control in young lambs. Afflicted lambs are born normally, but after a few days lose proper use of their limbs, particularly the hind ones. Some can drag themselves along, and a kind dam will stand over them and allow suckling. It is a distressing sight, and can be prevented by injecting the ewes with a copper preparation supplied by the vet around mid-pregnancy. It is worse in some years and some areas than others; your vet will advise.

Orf

Orf is a condition of older lambs, and sometimes ewes which develop unsightly lesions on the face and lips, making stock unsaleable as stores or for breeding. Orfoids given orally, or terramycin spray on the lesions, usually gives a cure, but far better 'scratch' the lambs with a preventative from the vet. His advice is essential, for if the condition is not orf, it will be after using the live vaccine. The disease is difficult to eliminate once it appears on a farm.

Footrot

Footrot brings lameness, and should be suspected as soon as a sheep hobbles. It is caused by an organism that can live in the soil for at least a fortnight. It brings on sore places between the claws or at the side of the hoof, and is of permanent concern to all flockmasters.

No sheep suffering from footrot will thrive. Ewes do not milk properly, lambs will not fatten. And it spreads rapidly, especially on wet land which, though not being a direct cause, softens the hoof. There is no footrot in Iceland, where I have seen sheep sheds with a muddy access that would horrify British shepherds.

Vaccination is now available, but expensive. A competent shepherd checks feet at certain specific times, and pares off surplus hoof. Individual cases are treated by a spray, or a paste containing copper sulphate, but a regular walk through a footbath containing copper sulphate solution or formalin is recommended.

Pregnancy Toxaemia

As lambing approaches, the nutritional needs of the ewe increase while her stomach capacity is lessened by the unborn lamb's growth. If she carries two or more, the problem increases. At a certain point the deficiency may become so acute that the sheep falls into a stupor or semi-coma, and several animals may be affected, as if an infection had hit the flock.

The answer is to feed less roughage, and that only of the highest quality, and ample easily digested concentrates. This usually works in a shed, but if the ewes are outside another factor enters the equation. Bad weather, notably driving rain, causes further energy loss from the ewe, who then goes down. In that case feeding glucose solution, or injecting glucose into a vein, may help. Anabolic steroids from the vet are now being more widely used.

Hypomagnesaemia (Grass Staggers)

This is preventable in theory, as it is simply a deficiency of magnesium and/or calcium in the blood stream. The animal's body cannot store magnesium; requirements must be met daily. Only a trace is needed but, particularly when there is a sudden flush of grass, that trace may be absent.

The sheep collapses, with glazed eyes, and may throw a 'staggers fit'. If caught in time, 60 to 80cc of calcium burogluconate injected under the skin in different parts of the body may save the situation. If attacks occur, carry bottle and syringe on every round; by the time you have reached the veterinary cupboard and returned to the field it may be too late.

Calcined magnesite is spread on dairy pastures to combat the problem, but you can't do that on a hill. Mineral licks should be offered, but there is no certainty that every sheep partakes. They won't eat magnesium from choice; it must be disguised, and liquid feeds mixed with molasses are a suitable carrier. High-magnesium concentrates may be fed right into June, and certainly help, but it goes against the grain to pay for cake when the pasture is shooting ahead.

Drenching gun.

Attachments for the drenching gun.

Draft hill ewes put on rich spring pasture can drop like flies from grass staggers. One practical point is never to use a compound fertiliser on sheep pasture in spring; leave the potash element till midsummer.

OTHER DISEASES

Foot and Mouth

Here's one we never wish to see. Foot and mouth disease is notifiable, the symptoms being high temperature (normal is 104–5°F, 40–40.6°C); lesions between the claws which cause lameness, and salivating. On the slightest suspicion, you *must* call your vet immediately, and of course a special watch must be taken if there is an epidemic.

Abortion

Few things are more distressing to the livestock breeder than an abortion 'storm'. All plans are set at nought, and a whole year's income seriously threatened. A lambing may start badly, with a few ewes lambing weak or dead short-term lambs, and then settle down. If more than the occasional ewe aborts in mid- or late-pregnancy, seek immediate professional advice with samples for the veterinary investigation, as there are a number of different types of infectious abortion. Ewes can abort through fright or rough handling.

Pneumonia

Vaccination is possible against this killer that can affect sheep indoors or outdoors; the economics are best discussed on the spot. Just as it is possible to over-insure, so a balance must be struck between cost of the various preventatives and probable losses. But peace of mind is worth a lot.

At the first sign of abnormal breathing, a penicillin injection should be given, 6cc for an average weight adult sheep, followed by 4cc on the next two days. Pneumonia can occur on hot days, especially if the nights are cold, and in soaking wet weather, particularly if it follows a dry spell and the sheep are unaccustomed to the change. Vaccination is possible, but is neither sure nor straightforward, because of the number of types of pneumonia.

Urinary Calculii

A blockage of the urinary tract can occur in wethers, particularly if housed on dry feed. There is no practical cure; send the affected animal to slaughter for what it will make, before it suffers more pain and the owner more loss. The carcass is quickly condemned, but part payment may be made if caught in the early stages. The last one we had became worse over Christmas when the slaughterhouse was not operating, needed a £6 veterinary certificate stating that no antibiotics had been used, and was then totally condemned.

Symptoms are lassitude and straining, and swelling around the wether's pizzle. Feeding soaked sugar beet pulp or roots does help as a preventative with housed stock. Adding salt to the drinking water to raise input is also recommended.

Lamb Diseases

WATERY MOUTH

Watery mouth is self-descriptive, and occurs in young lambs. It is an E.coli infection, and an oral antibiotic plus an antibiotic injection usually works.

COCCIDIOSIS

Coccidiosis comes with intensification. It is treated by medication, but is so serious that veterinary advice should be sought.

JOINT ILL

Joint ill will arise if navels are not treated immediately after birth with iodine or antiseptic spray. Lambs become lame and stiff-jointed.

8 Show and Sale

SHOWING

Some regard shows as a waste of time, or merely social events, bearing no relation to the commercial world, and too costly. While farming many people pass through that sort of phase, as I did, but they usually realise that the benefits of showing far outweigh the drawbacks. The first is that one's own stock is judged independently against fellow breeders', and that can be a salutary experience.

We can all work hard at home, tending our own stock night and morning, until we come to regard them as the ultimate. That hill ewe with bonny twins for the third year running – where can you find another to match her? The answer probably lies at the local show, and almost certainly at the county show. The show ring soon brings home deficiencies, and a fresh start in the right direction can then be made.

Show judges set standards for future generations. If a Welsh Mountain ram with a pronounced Roman nose wins the breed supreme at the Royal Welsh, others will be bred to try and topple him. Those with the desired characteristic will sell for more money, and the breeder must provide what the market demands.

The term 'job satisfaction' is a 1980s addition to the industrial vocabulary. Stockbreeders discovered it centuries ago, and the stimulus of turning out an animal that will please the judges and the crowds is an intangible asset that lightens the rather wearing routine that livestock imposes seven days a week.

For the pedigree breeder, publicity is essential. If his stock win, they may be photographed in the local and farming press. If they are entered but do not win, they may still attract the attention of someone seeking rather different qualities from the judge. And agricultural correspondents may mention them in the county newspaper, which they cannot do if not entered.

Showing often begins with no more than a genuine desire to help a local function. No matter; once having experienced the excitement and comradeship that exists among stockmen, other attempts follow, with incentive to use the best blood and manage the whole flock rather better. That is why showing elevates all standards.

SHOWING LOWLAND SHEEP

Dipping

An essential to show preparation is correct and timely dipping. Several dippings before the show season starts are advocated, and if only one or two sheep are involved, it is better to use a watering can than do nothing. A mixture of Show Dip and Blue Label is one recommendation for Down breeds; proprietary dips are available for imparting curl to the wool in those breeds demanding it.

Soft water or water that has been softened is advocated, and a breezy warm day is preferable to hot sun. Ideally the weather will be such that the sheep dry within three hours, and thereafter have shelter from heavy rain after the final dip before showing.

As these tips apply equally to preparations for sale, they have wide application.

Trimming

'It's not what you cut off; it's what you leave on that counts' is the trimmer's golden rule. Hand shears are generally used for show or sale preparation, with heavy ones for the early stages, and lighter ones for the final trimming.

Trimming should begin four or preferably six weeks before the show date. The first step is to cut the backs level, leaving on a quarter of an inch of wool, but more if the show date is nearer.

Have in your mind the ideal of a perfect sheep of the breed concerned. Then try to make each sheep approach that ideal by judicious clipping. Lessons from a fellow exhibitor are invaluable, and the skills learned may be applied to the whole flock. The best trimmers are true artists; they can create optical illusions out of moderate material, and unbeatable ones from the best.

Carding

The next stage is carding, which is preceded by a dipping after the final trimming. A carding board has masses of fine wires with turned tips which catch the wool and 'lift' it. Fairly heavy boards with a slightly convex curve are advocated.

For patting or 'scratching' the fleece, a flat card is used, with a bevelled card for the lifting process. Dirt is removed from the fleece by scratching down, and then lifted in a circular action. After damping down, the lifted wool is clipped off. When lifting sweep *against* the fall of the wool, and finally clean the carder using an old dinner fork with prongs bent at right angles.

A Y-shaped yoke on a stand is used to keep the sheep's head in place for clipping, which cannot be done properly if a halter is worn.

Use of the 'card' for scratching or patting the fleece level. Used on shortwools before show or sale.

Numbering

Stock entered for a pedigree sale will be allocated a catalogue number, and the same applies to shows. These numbers are stamped on the back of a shortwool, obviously after clipping is completed, and is one of the last jobs to be done before setting out.

Red cellulose paint is a favourite, and marking irons should have a round face for sheep in wool, and a flat one for close-clipped animals. These stamped numbers must correspond to catalogue numbers, and with ear tags.

It is surprisingly easy to stamp the wrong sheep or number if care is not taken. Stamping must be done precisely; haphazard, slapping-on of red paint looks most unprofessional.

The branding irons should be set out in numerical order, and the paint pot near them, safe from escaping sheep.

*Neat marking enhances an animal, whether for show or sale.
Hill Cheviot rams at Hawick.*

HILL SHEEP PREPARATION

As befits stock kept in 'natural' conditions, hill sheep receive less preparation for show. Trimming round the head and neck is permissible, with one or more dippings. The tail end must of course be kept clean, and there is always the danger that hill sheep run on better grazing for the summer shows will 'scour' on the lush pasture.

Selection

Where ram lambs are retained for show or sale, at least three times the required number will be left uncastrated. During the summer these are gradually whittled down, the process continuing during the winter if shearlings are to be sold.

The Scottish Blackface has spoilt itself by allowing ram breeding on much kinder farms than those where the animals must spend their working lives. Other hill breeds have adopted a more commonsense attitude, but high concentrate feeding of ram lambs is creeping in. Stick dressers say that the horn from such sheep is soft, and it follows that the same will apply to the bone. So undue cake feeding of hill breeding stock is unlikely to engender a reputation for soundness.

Wintering

Hill ram lambs are generally wintered in batches on their own, and given sufficient food to express their full potential, rather than being forced beyond it. A sheltered meadow, with perhaps a barn to run in and out, sweet hay and a little concentrate is the best recipe.

Summering

In May the tup hoggs return to the hill, but are clipped before the main flock, probably in June. That gives fleeces a little more time to grow before summer shows or autumn sales. Clipping is done carefully and not too short. Then the shearling rams return to the hill, when those with shortcomings revealed by the shears may become wethers.

Rams to be shown during the summer are kept handy, nearer home. The biggest danger to the rest is an attack of sheep head fly in districts where that pest is prevalent, and I have seen Swaledale rams completely disfigured when a group of them in a high allotment were not seen for several days during a busy haytime.

Autumn

During the weeks before the autumn sales, hill shearling rams have more human contact than at any time in their lives. Such a short period does not affect their hardiness, but it is an insurance that valuable stock will be shown to best effect.

Mid September to late October is the date for most ram sales, and shortening days bring greater risk of escapes and fighting. Fences must be foolproof, and sale rams are often housed each night as the date approaches. All sheep look better with a dry fleece, and the extra handling and confinement help accustom the animal to the sale ring. About ½lb (227 grams) of concentrates per sheep per day may be fed at

A sea of white Welsh Halfbreds at one of their well-organised sales.

this stage, but a mixture of bran, sugar beet pulp and cereals is more suitable than a high protein diet that may cause looseness. Any animal unduly loose in its dung is not converting its food properly.

Feet must be watched at this stage. Rams have an infuriating habit of walking sound for months, and suddenly succumbing to foot rot a few days before sale or tupping time. A regular run through the foot bath or a terramycin spray on every foot (not just the doubtful ones) should be routine. The feet are trimmed as required.

Dipping

A proprietary shampoo is given a fortnight before the sale, on a brisk day if possible, this shampoo being well scrubbed in by hand. Some shepherds add a bag of peat to the dip, especially where draft ewes are being sold, as this lends uniformity and gives an impression of high grazings. A red clay solution combats the slight blueness associated with certain peat grazings.

Halters are of leather, nylon or cotton. Each breed has its own fashion between leather or white cotton, the latter being more expensive and more satisfactory than nylon. Synthetics and livestock do not mix too well, and cotton is more comfortable.

Dress

Having spent hours preparing his sheep for the judge's eye, the exhibitor must be neat and tidy himself. If white coats are the rule, see that they really are white, and footwear cleaned to match. Slipshod dress on the exhibitor's part is tantamount to letting the animal down, and while in no sense on a fashion parade, the shepherd must look the part.

IN THE RING

A well-cared for appearance impresses buyers. A ram's genes will not be altered because his face is clean; but care in that department indicates care at other levels. A final face wipe adds to the smart appearance of a dark-headed sheep, while a chalk block lends a glistening white to

*Catching a sheep. These Swaledale rams are powerful
animals, but the horns are an aid. Icelandic farmers refuse to
breed the horns off their sheep owing to their value in handling
and branding.*

Border Leicesters and others with white faces. The blocks are sold at
dog shows or pet shops.

Rams of lowlands breeds may be easily trained to stand quietly and to
advantage, i.e. with forelegs perpendicular and parallel, hind legs
slightly back. Hill rams are not expected to be docile; they must lead the
flock in a storm, and a bright and alert eye is of more value than
placidity.

Horned breeds, notably Scottish Blackface, may have their horns
polished. Glass and varying grades of sandpaper are used, with a final
oiling. I'm never too sure whether this practice is really in line with the
character of a hill sheep but, if judges and buyers approve it, the
exhibitor should follow suit.

Show and Sale

A crop of rape for sheep feed on the Border hills.

Feeds

The shepherd does not usually indulge in 'secret' recipes, but food for shows must be considered, and especially at the bigger shows of more than one day. Ascertain whether or not green fodder is provided.

Growing Special Crops

A succulent, easily-grown green feed is needed for summer shows, and nothing beats the traditional oats and vetch mixture. Seed is now very dear, but only a small area is needed, and compared with the other expenses of showing this one may be shouldered lightly. Rye and ryegrass will provide early summer material, but are not quite as good. Mangels are an excellent feed, and will keep till July if properly stored in a cool building insulated under piles of straw.

For summer and autumn, cabbages are unsurpassed. A great weight per acre may be grown, and the crop keeps fresh for a week or more after cutting. In fact, specialists prefer cabbages at least a day old to those freshly cut. Sweet hay is an ideal accompaniment at all seasons.

Glossary

GENERAL TERMS

Allotment Walled enclosure, usually on a hillside, and of better quality than the hill above.

Barren, eild, geld A ewe not in lamb.

Broken-mouthed Older sheep with loose or missing teeth, unable to graze properly and therefore requiring easily-grazed herbage.

Cast An old sheep fit only for the butcher. In some districts, a sheep fast on its back unable to rise.

Couped, rig-welted A sheep turned onto its back and helpless.

Crone Old ewe.

Crop Preceded by a number, denotes the number of lamb crops that the ewe has had.

Draft ewe At four to six years old, the hill ewe is drafted downhill for further breeding.

Ewe Adult female.

Fell, hill, moor Regional names for upland areas of natural grazing, sometimes unfenced.

Gimmer Young female. A gimmer lamb is a ewe lamb.

Heft An area of hill grazed by certain sheep, who become 'hefted' to that part.

Hirsel   A number of hefts, comprising the area of ground herded by one shepherd, or the sheep on them.

Hogg, teg (also hogget, hoggerel)   From December of the sheep's first year to its first shearing the following summer.

Inbye   The fenced or enclosed lower fields of a hill farm.

Lamb   Young sheep, from birth to weaning or the end of its year of birth.

Lunky hole, smout hole   Specially made low hole in a stone wall; takes only one sheep at a time, and is easily blocked when not in use.

Polled   Hornless.

Ram, tup   Interchangeable terms for an uncastrated male.

Rud, raddle   Coloured powder, the base of a marking paste.

Spain, wean   The act of taking young stock from their dams. End of the milk-feeding period.

Strike, struck   Shepherds' terms for attacks of the blow fly, maggots from whose eggs attack the living flesh.

Terminal Sire   The one used in the final cross for fat lamb production. All his offspring, male and female, are graded.

Theave   As gimmer.

Twinter   Sheep in their second winter.

Two-shear   Sheep in its third year. Most are first shorn at just over one year old.

Glossary

Two-tooth   As gimmer.

Wether   Castrated male for eventual fattening.

WOOL TERMS

From Michael Ryder's *Sheep and Wool for Handicraft Workers*.

Break   Tender part of staple.

Britch or breech   Coarse wool from hind legs.

Cast   Fleece that does not reach grade standard.

Character   Regularity and definition of crimp.

Count   Degree of fineness (*see* Quality).

Crimp   Natural waviness of wool fibres.

Crossbred   Wool less than 60's quality, not necessarily from crossbred sheep.

Daggings, doddings   Wool from around the tail area, containing solid lumps of dung.

Dead wool   Misleading term to describe hairy fibres; all wool is dead!

Deep   Term applied to long wool.

Density   Closeness of fibres to one another in the staple.

Grade   Standard on which wool is bought and sold.

Grease   The skin secretion lanolin.

Handle   Degree of softness.

Kemp Short, very coarse fibres. Appear chalky white and have pointed tip and tapering root.

Quality, count, number Degree of fineness.

Rise Spurt of wool growth in spring.

Staple Natural lock or tuft of fibres.

Strong Coarse wool of low quality.

Suint Deposit on fleece from sweat gland.

Woollen Yarn made from carded wool.

Worsted Yarn made from wool that is combed after carding to make the fibres parallel.

Yolk Yellow discoloration of fleece caused by mixture of grease and suint.

Further Reading

The Ark (monthly), Rare Breeds Survival Trust, Winkleigh, Devon.

Bowen, Godfrey, *Wool Away!* (Whitcombe and Tombs, 1955).

British Sheep, (National Sheep Association, 1983).

Cooper and Thomas, *Profitable Sheep Farming* (Farming Press, 1972).

Fell, Henry, *Intensive Sheep Management* (Farming Press, 1979).

Firbank, T., *I Bought a Mountain* (Harrap, 1940).

Fraser, Allen, *Sheep Farming* (Crosby Lockwood, 1937).

Grant, D., and Hart, E. *Shepherds' Crooks and Walking Sticks* (Dalesman, 1972).

Hart, Edward, *Northcountry Farm Animals* (Dalesman, 1976).

Hart, Edward, *Showing Livestock* (David and Charles, 1979).

Hart, Edward, *The Hill Shepherd* (David and Charles, 1977).

Longton T., and Hart, E., *The Sheep Dog: Its Work and Training* (David and Charles, 1976).

Morrison, A.M., *Red Dragon Farm* (Faber, 1964).

Nixon, David, *Walk Soft in the Fold* (Chatto and Windus, 1977).

Perry, Richard, *I Went A-Shepherding* (Lindsay Drummond, 1944).

Rainsford-Hannay, Col. F., *Dry Stone Walling* (Faber, 1957).

Raistrick, Arthur, *The Pennine Walls* (Dalesman, 1972).

Rollinson, William, *Lakeland Walls* (Dalesman, 1972).

Ryder, Michael, *Sheep and Wool for Handicraft Workers* (Available from 23 Swanston Place, Edinburgh).

Useful Addresses

Animal Breeding Research Organisation, West Mains Road, Edinburgh.

British Wool Marketing Board, Oak Mills, Clayton, Bradford, W. Yorks.

Country Landowners' Association, 16 Belgrave Square, London SW 1.

Grassland Research Institute, Hurley, Maidenhead, Berkshire.

Hill Farming Research Organisation, Bush Estate, Penicuik, Midlothian.

Institute of Animal Physiology, Babraham, Cambridge.

International Sheep Dog Society, Bedford.

International Wool Secretariat, Wool House, Carlton Gardens, London SW 1.

Joint Sheep Group, National Farmers' Union, Agriculture House, Knightsbridge, London SW 1.

Meat and Livestock Commission, Queensway House, Bletchley, Milton Keynes.

National Institute of Agricultural Botany, Huntingdon Road, Cambridge.

National Sheep Association, Jenkins Lane, St. Leonards, Tring, Hertfordshire.

Index

Page references for illustrations are indicated by italic type.

Abortion 100
ABRO 12
Anglesey 21
Arable 1, 17
Artificial insemination *42*
Ashford 20
ATB 49
Autumn sales 4, 107

Barren ewes 36, 48
Beulah Speckle Face 6, 30, 32
Black Welsh Mountain 33
Bleu de Maine 22
Blue-faced Leicester 6, 8, 16, 17, 22, 29, 30, 55, 73
Bloom dips 74
Boaz, Geoffrey 12
Body scoring 38 – 40, 45
Border Leicester 10, 22–3, 24, 30, 32, 110
Boreray 33
Bowen, Godfrey 65
Brecknock Cheviot 29, 30
Breeding cycle 40
Breeding stock 3, 6, 13, 34
Breeds, choice of 16
British Friesland 13, *13*, 14
British Milksheep 24
Builth Wells 18, 19, 30

Cabbage 111
Calendar 34
Carcass 15, 18
Carding 104, *104*

Care of ewes 43
Cash flow 3
Castration 57–62, *60*, *61*
Charollais 22
Cheese 3, 13–14
Cirencester, Royal College 19
Clark, Hugh 20
Climate 16
Clipping 10, 104, 106
Clun Forest *2*, 17, 23, 32
Coccidiosis 101
Colbred 24
Colostrum 14, 55
Concentrates 41, 45– 6, 100, 107
Condition scoring 38 – 40, 45
Copper deficiency 4, 6, 16
Copper poisoning 14
Cotswold 33
Creep feed *57*, *59*
Crutching 21, 40, 47
Cull ewes 11

Dairying 17, 20
Dalesbred 6, 23, 26, 29
Dartmoor 33
Derbyshire Gritstone 27
Devon Closewool 33
Devon & Cornwall Longwool 33
Devon Longwool 33
Dipping 21, 63, 70–5, *72*, *73*, *74*, 89, 103, 106, 108
Diseases 91–2, 96, 97
Dogs 12, 27, 43
Dorset Down 15, 18

118

Dorset Horn 3, 18, 34
Dosing 94, *99*
Down breeds 17–24
Down rams 8, 10, 11, 24, 29, 32, 34, 37
Draft ewes 4, 6, 9, 24, 25, 26, 36, 63
Drenching (see dosing)
Dress for showing 108
Dutch East Friesland 24

Early fat lamb 3–4
EEC 4
Electric fencing 79, *79*
English Halfbred 23, 32
Equipment 76
Ewe lambs, breeding from 3
Ewe management 38
Exmoor Horn 10, 33

Fat lamb 3, 4, *5*, 6, 11, 22, 25, 37
Feeding at shows 111
Feet 38, 40, 47, 108
Fell, Henry 21
Fencing 76–80, *80*, *81*
Findon 20
Fleece 11, 13, 18, 23, 26, 27, 33, 37, 45,
   62, 63, 93
Flock mark 62
Foetus 43, 44
Foot-and-mouth 100
Footbath 38, *88*
Footrot 97–8
Fraser, Dr Allan 44
Fries Melkschaap 62
Fullwood machines 14

Geld ewes 43, 47
Gestation 40, 46
'Gimmering' 3, 10–11
Grader 5, 6
Grassland sheep 32
Grass staggers 99, 101
Great Yorkshire Show 19
Guarantee 15
Guide price 15, 17

Halfbred 3, 4, 6, 9, 11, 17, 18, 20, 22, 24,
   26
Halter 108
Hampshire 15, 18
Handling yards 89–90
Harness 40, 43
Health 36, 91
Hebridean 10, 33
Hedge laying *83*
Herdwick 27, *28*, 67
HFRO 38
Hill Cheviot 9, 23, 24–5, *25*, 27, 30, *105*
Hill Radnor 32
Hill sheep 7, 16, 22, 24, 35, 37, 45,
   106–8, 109
Horn burning 62
Housing 4, 84–9
Hypomagnesaemia 99
Hypothermia 54–6

Innoculations 46–7, 91, *92*, *93*, 100, 101

Jacob 33
Jameson, Harry 62
Joint ill 101

Kelso 18, 19, *25*
Kendal Masham 23
Kerry Hill 32

Lambing 34, 44–62, *50*, *53*
Lambing equipment 48–9, *58*
Lambing percentage 4, 14, 30
Lamb diseases 101
Leicester 10, 33
Leys 1
Lincoln 10, 33
Livestock farming 29
Liveweight gain 21
Llanwenog 32
Lleyn 32
Longwools 33, 63
Lonk 6, 27
Lowland sheep 15, 16
Lustre wool 23, 33

Maidstone 20
Malton 19
Mangels 112
Marking 62, 105
Manx Loghtan 33
Masham *12*, 17, 23, *23*, 26, 29, *31*, *42*
Meatlinc 21
Meatline 21
Merino 11–12
Metcalfe, Ernest 70
Mid-season fat lamb 3, 4–5
Milk 3, 13–14
Milk substitute 14, 35
Mineral licks 46, 98
Mispresentations 51–2
MLC 2, 15, 39, 93, 96
Moredun Research Institute 55
Mothering on 56–7
Mothering up 69
Mule 9, 17, 22, 25, 29, 34, *42*, *71*

National Lambing Competitions 29
Netting 76
New Zealand lamb 20
NFU 49
Northampton 19, 21
Northcountry Cheviot 6, 23, 24, 30
North of England Mule Sheep
    Association 30
North Ronaldsay 33
NSA 2, 27
Numbering *54*, *59*, 105, *105*

Oestrus 3
Oldenberg 22
Orf 97
Owen, Iolo 21
Oxford 15, 19, *20*

Pasture 4
Pedigree 3, 6, 10, 34, 102
Pelts 3, 11–12
Pneumonia 100–1
Polled Dorset 3, *8*, 18, 34
Portland 33

Pregnancy 43, 45
Pregnancy toxaemia 98
Production costs 4
Profit factors 14–15
Prolapse 36, 50
Progeny test 6, 29
Pure breeding 37

Quality 36

Raddle 40, 41, 46
Radnor 6
Rainfall 16
Ram breed comparison 15
Ram breeding 36
Ram feeding 41
Ram management 31
Ram selection 36–7
Rape *110*
Rare breeds 3, 10, 33
Re-absorption 43
Recording 62
Redesdale Experimental Husbandry
    Farm 9, 29, 62, *87*
Richardson, Bill *86*
Ridley Rappa 79–80, *81*, *88*
Romney 16
Rough Fell 6, 23, 26–7, 29, 67, *95*
Royal Show 18–19
Royal Welsh 102
Russel, Kenneth 19
Rumenco 29
Ryeland 17
Rye and ryegrass 111
Ryder, Michael *12*, 114

Scanning 47–8, *47*
Scots Halfbred 23, 24, 30, *32*
Scottish Blackface 6, 9, 17, 22, 23, 25,
    27, 29, 30, 67, *85*, 106, 109
Scottish Colleges 2, 55
Scottish Greyface 23, 30
Scrapie 37, 96
Semen testing 38
Shearing 21, 63–9, *66*, *67*, *68*, *70*, *71*, *86*

Sheep scab 96
Sheep Systems Promotion 22
Shepherd of the Year 8, 29
Shetland 33
Showing 102–5
Show dipping 73
Shropshire 18–19
Silage 43, 45, 45
Smallholder 1, 6, 12, 17, 18, 33, 37
Soay 33
Soil 16
South Devon 33
Southdown 15, 20
Stick dressing 106
Stocking rates 1, 14
Stone walls 82, 83
Store lambs 3, 5, 6, 9, 11, 25
Sugar beet 11, 45
Suffolk 3, 12, 15, 18, 19, 21, 24, 27, 57, 59
Swaledale 6, 9, 17, 22, 23, 25–6, 26, 27, 29, 35, 37, 45, 52, 109
Swayback 46, 97
Swedes 11

Tailing 41, 60
Taunton 18
Teeswater 6, 22, 23, 23, 29
Teeth 35–6, 35
Terminal sires 21–2
Texel 15, 21, 22
Thomas, Jack 8
Trace elements 16

Trimming 103
Troughs 89
Tupping 4, 40–3, 108
Turnips 11

Udders 34–5, 38, 47
Universal Sheep Cradle 88
Urinary calculii 101

Vaccination 40, 46, 76, 91
Vasectomised rams 4
Ventilation 84, 93

Watery mouth 101
Welsh Badger-faced 33
Welsh College 2
Welsh Halfbred 23, 30, 31, 107
Welsh Hill Speckled Face 29, 30
Welsh Mountain 6, 16, 23, 27, 28, 29, 30, 52, 102
Welsh Mule 30
Welshpool 30
Wensleydale 6, 22, 23, 29
Wether 9, 11
White-faced Woodland 33
Wiltshire Horn 21
Winter dip 70
Wool 3, 11–12, 33, 38, 63
Wool faults 68
Wool terms 114
Worms and worming 1, 40, 94–5
Wrapping fleeces 63, 66–7